On Being a
CONCEPTUAL ANIMAL

Arthur Niehoff

The Hominid Press • Bonsall, California

Copyright 1998 Arthur Niehoff. All rights reserved
ISBN 0-9643072-4-3
Library of Congress Catalog No. 97-76969

Book typography: Christa Simonsen
Cover: Robert Howard

Manufactured in the United States of America

1st Printing

Publisher's Cataloguing in Publication
(Prepared by Quality Books, Inc.)

Niehoff, Arthur H., 1921-
    On Being a Conceptual Animal / Arthur H. Niehoff. --
1st ed.
    p. cm.
    Preassigned LCCN: 97-76969
    ISBN: 0-9643072-4-3

    1. Anthropology. 2. Concepts. 3. Niehoff, Arthur H.,
1921- I. Title.

GN31.2.N54 1998   153.23
                          QBI97-41429

Table of Contents

Introduction ii
1.) Concepts to Live By 1
2.) The Pleasure of It's Being 27
3.) On Becoming a Professor 39
4.) Dealing at the Convention 54
5.) The Publishing Game 65
6.) The Sorcery Book Publishers 73
7.) In Praise of Libraries 78
8.) From the Top of a Hill 87
9.) Curtains 94
10.) On Sorting Garbage 101
11.) The Non-Endangered Species 114
12.) The Glorious Sun 126
13.) From Cow Slaughter to Ahimsa 137
14.) Is Peanut Butter Food 148
15.) Love Affair with Avocados 154
16.) Us and Them 163
17.) The Evolution of Presenting Behavior 183
18.) On Being a Barnyard Watcher 195
19.) Love Magic 210
20.) The Real Life Church 218
21.) A Tall Tale From the West 231
22.) The Stroke 243
23.) On Being a Latter Day Veteran 253
24.) On Becoming Older 267

On Being a Conceptual Animal

Introduction

We humans like to consider ourselves special, both as individuals and as members of groups. Thus, it should hardly come as a surprise that those who have dedicated their lives to the study of humankind as members of the species, anthropologists, should be very interested in human uniqueness. They have characterized *Homo sapiens* and his ancestors as creatures which stand upright, make tools, speak in languages, have large brains, and other characteristics.

The human creature is also thought of as a rational, thinking animal. I perceive a human being as a conceptual animal. It does not perceive the universe directly but through a system of concepts or ideas. In fact, I will venture that a creature without concepts is not human. A skeptical reader may quickly ask, "Are you saying that other animals do not think, do not have ideas?"

And my answer is, "Certainly not of the same kind and to the same extent." Here I will have to agree with the great anthropological philosopher, Teilhard de Chardin, that with man a qualitatively different thinking process has evolved, and which for better or worse is now the predominating force of the world. He thought it was a very progressive force. I am not so sure of that but I do believe that we are stuck with it.

The best we can do to characterize the other animals is to describe their behavior. Ethologists, persons who observe and describe animals in their natural habitats, characterize them by what they eat, carnivorous or herbivorous, or in what kind of social groupings they live in, multi-male or harem or other behavior. And here the questioning reader can ask, "But how does that tell you that chimpanzees or lions do not think?"

So far as we can make out, they do not think extensively or in a complex fashion. Teilhard claimed that they thought with instincts. But also of course they think with the learning they get from experience and imitation. Whether they are the pigeons, deliberately trained by Skinner, or deer late in the hunting season, these animals have learned how to get food pellets and to be very wary of bipeds with firearms, not a part of their genetic make-up.

Introduction

Admittedly, we cannot enter into the mind of a chimpanzee or lion, for one simple reason -- they do not have a language. There have been quite a number of attempts to teach the great apes to use human languages. But the results have not been clear. However, even if these animals did learn very well to react to human speech, that is far from the species having a language of its own. We know that all bisexual animals communicate, but it is communication of a simple sort, what we call by signs, rather than by symbols. Humans have evolved communication systems which depend on arbitrary meaning for the things and events of the world, and even to figments of the imagination. The units of arbitrary meaning are called words and our entire universe, both seen and unseen, has been so named. Thus we live in a world of word-symbols.

We cannot even get the necessities of life through direct thought, whatever that is. We always have to go through the system of symbols, whether in our own minds or through communication with others. We have names for everything, including the food we eat, the work we do, and the sex acts we perform. We do not simply have sex with a mate whose odor pleases us. Rather we have to go through a complex procedure, which in Western culture is called dating. This may lead to getting together in a room where there is another series of prescribed procedures. Then if everything is satisfactory to both parties, we may get to what we are after, following the use of a tremendous number of words. We are indeed word animals and the word is the essence of an idea.

Most of us were early taught that we are special in being rational. We were supposed to dispassionately analyze world and life events and come to "intelligent" conclusions. That was thought to make us superior to other animals. This idea is not so popular nowadays, after the era of Freud and Darwin and the study of anthropology. We now generally concede that humans also have many ideas, and act accordingly, because of emotion or other feelings.

Many of us have ideas based on revelation. Thus we believe that our spirit deity rose from the dead after dying on the cross or that the world was created by a water creature who brought successive mouthfuls of mud to the surface to build up the land areas. But no matter where our ideas come from, if they persist through time, they must be embodied in words. Many philosophers, at least as far back as the ancient Greeks, have considered the significance of ideas as embodied in words. Having a background in anthropology, I have more familiarity with man as an animal and how this has affected his behavior.

Anthropologists believe that *Homo sapiens*, our species, differs from other species in that he is the creature who stands upright and has the largest and most complex brain proportionate to his size. Having his hands free enabled him to make and use tools, as well as to carry many objects freely. Also, the large brain enabled him to cogitate extensively and to evolve a language.

We especially think of him as a tool making animal. But what is a tool anyway? Is it a piece of stone that has been chipped into a spear point, or an assembly of metal and plastic which can carry us from one place to another on the surface of the earth. I will suggest that it is the idea of those things more than the things themselves. Whenever anything was invented by a human being, it was given a name, both whole and in part, along with the methods for making it. Otherwise, how could the inventor tell somebody else how to make or use it? How could the spear point or the jet airplane be made and used for generation after generation?

There was a not very bright saying by a military spokesman during the Vietnamese War, that if those people continued fighting, "We would bomb them back to the stone age." His thought was based on the belief that the reality of a culture is the physical things of which it consists. I offer that it is not the things at all, but the idea of them which is the ultimate reality. The only way the Vietnamese could have been bombed back into primitivity would have been to destroy all the ideas. That would have to include all the engineers and other builders and all the instruction manuals, both in and out of the country. Because they or any other peoples could have quickly rebuilt everything. The German and Japanese "miracles" should have taught everyone this. In short, ideas are the absolute basis of everything that makes up humankind's culture.

So change takes place in a culture when there are new ideas which are inevitably embodied in words. There is an old saying in English, and probably in other languages, that "experience is the best teacher." I will offer that this is not true unless new concepts are considered as part of experience. Most old people are not much wiser

Introduction

at the ends of their lives than they were at the beginning. They have had much experience, but without new concepts, they could hardly use it. The most natural thing to do is to continue what one has been doing.

Learning and using new concepts requires change in thought and behavior. Thus most people resist eating new foods during most of their lives. As infants they begin as blank slates with no biological imperatives except to nurse. Then they are offered some of the food of their culture which they will tend to lock onto by the age of three or four. It will be rice, soy sauce and pork for Chinese, hamburgers, french fries and catsup for North Americans. In their late teens they may come to know some other foods, and even try some, the Chinese trying french fries, the North Americans rice. As adults, they may continue to experiment, though within limits. The traditional Chinese person will not drink cold milk with a meal while the American will not eat dog meat. And when they get into their later years, they will become less and less innovative, preferring the food of their culture. They are used to it. Habits are not easily changed, because to do so requires getting and using new ideas.

Where do all these new ideas come from? Does each of us go around generating such all the time? As individuals we do come up with new ideas on occasion, probably most often ones that have already occurred to someone else. But most new ideas have not been dreamed up on our own. They are the products of our culture and we have learned them from birth on, especially after we learned our language. We then learned what is real food. As Westerners, we learned that ground beef and cold milk are food but that dog meat and insect larvae are not, even though all those items are nutritionally worthwhile. And as I will describe in the essay, "Is Peanut Butter Food?" many Latinos who lived in villages do not consider peanut butter food because they were never served it that way when they were growing up. The North American got used to it in his school lunch and continues to eat it. I will offer then the idea that "Foods are substances which are nutritionally beneficial and approved of in at least one culture."

And so in the process of brain washing we call enculturation we get multitudes of ideas for conducting our lives. They make up the patterned wisdom of our group. But that is not all. Groups change, and when they do, they invent new words. Then others borrow these new things and new words. Cultures borrow just as individuals do, and the primary way is from other groups. It is called diffusion.

What is the source of all the new ideas? To the individual, apart from the few he generates on his own, they come from history, revelation, science and other sources. We have in American culture a generalization that "all men are created equal." This comes from our founding fathers, who arrived at this conclusion from studies of earlier political systems. Also many of us have the idea that God is a superhuman being, embodying three entities in one. Buddhism is based on the idea that suffering is inherent in existence and based on the desire for material goods. And we have the idea that all living forms have evolved from others through time. This last, Darwinism, is one of the bedrocks of the idea system, science, with which I am most comfortable and which I will depend on in this book. Also I will rely on the ideas of my lifetime preoccupation, anthropology. However, I wish to emphasize here and elsewhere that important ideas can come from any thought system. It is merely that I got caught up in the web of anthropology and now rely on that system of ideas.

And what shall we do with these complexes of ideas? I gave this question much thought after I had written most of the essays that follow. I was then nearing the age of retirement, when it was no longer necessary to use my field of study for any practical purpose. Like the practitioners in some other academic fields, we anthropologists have been taught to do field studies, then to describe and analyze what we have seen. I have done this. However, late in my career I realized I had been observing and participating in events that were not standard field studies. Moreover I had been describing and analyzing them. They were the events of my daily life. It was participant observation of a complete sort, much more involved than the usual events of a field trip. I was using the concepts of my scholarly discipline to analyze my existence. I was not so much practicing anthropology as living it. And because I knew very early that concepts came from all fields of study, I thought that as an illustration

Introduction

of searching for, and utilizing, concepts for living, anthropology would do just as well as some other field of study about humankind. In a sense, it was life as a field trip.

From teaching, I am familiar with college students. Whatever their goal, they ostensibly go to college to get new ideas. Before they go, they are frequently asked what their academic major will be. They usually say that they will specialize in computer programming, nursing, police science, law or some other academic subject. It is much less frequent that one hears a young person say he/she is going to be in liberal studies, or general studies, or undeclared. I have never heard of a student who said, "I will major in ideas." Perhaps a philosophy student might think that, but even he would more likely use the formal subject title, philosophy. Furthermore, I think that members of the general public are uncomfortable when they are told by a young person that he is specializing in philosophy or anthropology or one of the other "non-practical" fields of study. To be amiable, the older person might say, "I suppose you plan to teach it, no?"

Or the questioner might be bolder and say, "But what will you do with it?" Meaning of course, "What job will you get?"

The usual American will feel more comfortable if the student says, "Engineering, or nursing, or medicine or special education."

The questioner can then visualize the occupation that will result.

Americans are particularly known as a practical people, interested in results. And in general our current students reflect that characteristic, they are usually "practical" in their choice of college studies. And as budgets get tighter, the universities keep pushing harder toward the same goal. Students are encouraged to judge their college careers almost exclusively in terms of the job at the end of the studies. Of what use are ideas for their own sake, most of us might ask? The prime goal of life is to achieve a certain standard of living, and for that a certain income is necessary.

And without denying the value of learning something that can be used to buy the bread, as well as the station wagon and house in the suburbs, is work all there is to human existence? As humans, aren't we organized physically, through blind evolution or purposeful creation, to achieve something more? Do we have to evaluate all the

On Being a Conceptual Animal

work we do according to its dollars and cents value? Shouldn't we be thinking of the quality of all our life as well as how much money we will make?

I shall illustrate this quandary from my own life, since I have had a wide range of experiences in this regard, and I think I have acquired a number of new ideas. I lived through a period when economic conditions were much worse than they are now. I grew up in the Depression years, when the standard of living had not just fallen, it had fallen very low.

Not only had wages dropped, one was lucky to have a job at all. For the most of us there was no question of maintaining the old standard of living; one was concerned to keep it from falling too far. And government assistance was much more limited. For most problems one was on one's own. When I got out of high school, the only job I could get was as a laborer in a factory. This was monotonous, dirty work on the night shift, for which I was paid $15 per week. I just could not stomach it so decided to retrain myself -- on my own, since there were no government programs for such. Like my parents, I then thought of education as being exclusively a matter of job training. So I entered myself into a business college which I paid for with my minimum savings. My parents had supported me through high school, but like most in the working class, they did not support me in education beyond that. They had barely enough to support themselves at that stage. Anyway I first learned shorthand, typing and simple accounting. I then got a job, keeping tally of sales in a meat packing plant for $15 per week, where I stayed until I got a job that paid a little more. I should stress here that I am not describing the economic conditions to show how much I suffered. Like most other young people of that time, I thought this was natural. Anyway, all this ended when I went into the army, as all us young men were required to do. This was not a very rewarding experience, nor did I do anything heroic. I was mostly a pawn of the military machine. I did almost get killed but was lucky enough to survive and to keep wondering if there wasn't something better in life. Probably the best thing that happened to me was to have been sent overseas for a while, though at that time I knew little enough about other places to get much out of what I experienced.

And what was the ideological component of those twenty odd

Introduction

years? Whatever I got must have come from reading on my own, a habit I picked up early in childhood and which I have continued ever since. My parents hardly encouraged independent reading, nor did the nuns in the Catholic school I was sent to. In fact, my independent reading was a source of conflict with both my parents and the teachers, since they were trying to keep me within the confines of Catholic ideology. I on the other hand was trying to spread my wings. Anyway, when my military career was about over, I was determined to go to college. I chose English as my major, because even then they were trying to push us into academic specializations. The only occupation I could think of was writing, but even that was vague in my mind. But in college I reveled in the new ideas I was exposed to, in all kinds of classes. Eventually I settled on anthropology as the most exciting and idea producing subject.

And this is a hard idea to get across to students nowadays --- from the day I started until a few months before I departed to take up my duties on my first professional job, I never thought about what I would do to make a living. I was too busy enjoying the multitude of new ideas. Moreover, I was not rolling in money. I got the payment from the G.I. Bill during most of those years, which as I recall was about $125 per month. My new wife worked at a low paying job as a nurse while I was an undergraduate, and I worked part-time as a secretary when I was a graduate student. By then we also had a child to support. Anyway, things improved financially slowly after that. I earned about $5,000 per year on my first job, and after that my salary improved steadily. When I got to be sixty-six, I retired comfortably, and my finances have improved ever since. During all this time I was working in anthropology and playing with ideas constantly. Now, insofar as my occupation and economic state are concerned, I consider myself very fortunate.

There is another way to consider playing with ideas, which is how we parcel out the time of our lives. In urban industrial culture we spend less than half of our life time as workers. The other parts we spend as students and in retirement. But since we only work about forty hours per week, we have 56 hours for sleep and 112 for doing other things. During our retirement years we have all our waking hours for other things. So what do we do with all that extra time? And as the reader can guess by now, I believe that if we search for and play around with ideas, we will be the most satisfied we can be as human beings.

Many parents may recognize the following as one of the most familiar complaints of children, "Aw gee, there's nothing to do." Of course there are multitudes of things to do! It is simply that the inexperienced child has not lived long enough to have acquired enough ideas to visualize many things to do. That will only come as the child acquires more and more interests, gets many new ideas. And to continue with the problem of childhood, consider the extension of the above complaint, "I'm so bored."

What causes a person to get bored anyway? It is because at the time of boredom the person does not have enough ideas to stimulate mental activity. Of course some situations make activity difficult, but I am perpetually amazed by the ingenuity of the human animal. Prison life has to be inherently boring. After all, it is supposed to be a punishment, and curtailing, or at least making difficult, any wanderings in the field of ideas must be as great a punishment as can be imagined. The ultimate punishment that men have dreamed up is solitary confinement, where the individual is cut off from almost all outside contacts. The potential boredom must be terrible. But many men have found ways to surmount even the boredom of prison life, practically always through the manipulation of ideas. Reading and writing must be the simplest way of evading boredom for them, since nothing is required but books, paper and pencils. And men have turned to that method. One of the most influential books I have read is <u>The Discovery of India</u> by Jawaharlal Nehru. It was written while he was imprisoned by the British for fomenting independence. Many American prisoners have assumed the occupation of jailhouse lawyer, immersing themselves in legalese. This kind of playing with ideas has the added advantage that it might help the prisoner get free. Some important legal precedents have been so established. One which quickly comes to mind is the Miranda Decision when a convict trained himself sufficiently to get court attention and bring about a repeal of his conviction. A precedent was established that no future convictions would be valid if the accused did not have legal counsel. I now have a nephew in prison who is passing his time studying to be a novelist.

So I would consider the human creature to be "the ideational

Introduction

animal," a consequence of potentially being "the boredom animal." Physiologically it is the complexity of the brain that causes this boredom. If it is not activated by ideational activity, it turns on itself, to create the condition of discomfort we call "being bored." The cure, of course is to become active in the manipulation of ideas.

And so to come back to the comparison of humankind with the other animals. So far as we can make out from observation, when their basic needs are fulfilled -- food, shelter, health and sex-- the other animals normally rest. In contrast, humans need mental activity or they get bored, even when the basics are fulfilled.

I still remember clearly a bizarre incident which occurred during my childhood. It was when the Kentuckians whittled the house down. It took place during the Depression years when many men were out of jobs. The Kentuckians were people who had come from the mountain region to my home town of Indianapolis to get factory jobs. There was little work in the hills and hollows where they came from. Still many of them had no jobs in the city either, like so many others. Many houses were empty also, since people did not have enough money to pay the rent. So a group of Kentuckian men would sit on the porch of an empty house at the end of my street and whittle pieces of wood that they would take from the siding. They whittled away steadily until the house was reduced to shavings. So far as I remember, no-one tried to stop them, I presume because the house was not that valuable and there were many others empty. And apart from the whittling, the Kentuckians talked and chewed tobacco, all activities essentially human.

Now how does this occurrence shed any light on the problem of boredom and the value of ideas? I will claim that humans, unlike other animals, find it hard to do nothing for long periods -- even if it is something as unproductive as making shavings out of boards. So, what about ideas, did those Kentuckians have any? And here I will venture that though they had no ideas for making anything out of their whittling, they were manipulating ideas in their verbal exchange, because that is what talk often is, even gossip. But the main point is that, unlike other animals, they were unwilling to simply just sit -- or lie.

The astute reader may guess by now that this is going to be a "how-to" book, which it is indeed. It isn't of course the standard variety. It will not help the reader lose weight, make a million dollars, have the perfect relationship, or one of the other ways to improve life, American style. But it will steer the reader toward the most human goal I can think of, to actively search and play with ideas. And only this will make him/her fully human. And though ideas can come from any source, in this book I will emphasize the way of science as the fount of thought. This is because that is what I learned and what has given me most satisfaction. And though learning, the acquisition of ideas, can take place in many different social settings, here I will emphasize what is learned in the institution of higher learning of contemporary society, the college. And again that is because that is what I am most familiar with. And finally it will be mostly the concepts of anthropology, for the same reason. Anyone who wishes to substitute any other source of ideas in thinking about my suggestions is of course welcome to do so. Because, remember I am merely emphasizing that the human is an ideological animal, but of all kinds of ideas. The Australian aboriginal who thinks of the origin of life as taking place during Dreamtime is being just as human as the biology professor who thinks that it is a matter of simpler creatures evolving into more complex ones during the last three billion years. Both are manipulating ideas that have been learned from others.

I originally thought of this piece as being aimed at beginning college students, and especially those who are concerned almost exclusively with learning as a means of job preparation. They will be the least concerned with learning ideas for their own sake, for improving the quality of their lives. And although my son ultimately chose the standard of living option, there was a time when he was an "idea" man. He had a perfect grade point average in college. One day he asked me whether I thought he should take a certain sociology class. He had heard that it was very interesting but that the professor was a hard grader. I told him that if he was primarily concerned with the grade point average, that he shouldn't take the course; but that if he was more interested in learning new ideas, that he should. He did, and got an "A." But to me what was most important was that he thought he learned a lot.

I have no illusions that I will change all the job learners into idea learners. But I have long given up the notion that I as an academic am going to convert students or readers completely. My

Introduction

only hope, then as an academic, and now as a writer, is that I can inject a small influence, plant the seed of an idea which may help to cause some to do some learning for the sheer human-ness of it. Sometimes when they are learning, they will exult in the idea, and forget the dollars and cents it will add up to. That's all I ever thought was possible in a classroom.

There is another kind of student who will probably find my message comprehensible from the beginning, the older, returning, experienced person. I have had many men as students who had reached early retirement but whose minds were very active, even restive. Many were already seekers, as I must have been when I entered college after military service. And there were many women who had either raised their children or were divorced and came back to finish the requirements for their degrees. Or they sought answers to life's puzzlements. Or they came simply to learn more about their universe. Then there have been the ex-Peace Corps Volunteers who came back to college because they learned in the field how much they did not know. I suspect all of the above will quickly understand what I am talking about.

Then even before I left the classroom I was thinking of that reader who enjoyed learning new ideas but who had never got to college or who was long gone from it. And this definitely includes the older person. This idea has been reinforced much as I have seen more and more learners in their later years. Many read voraciously. Older persons are frequent in the public library I go to. I witnessed an old lady some years ago who with the help of a younger man, perhaps her son, was trying to buy a magnifying glass for her failing eyesight. Her comment, which struck me deeply, was, "Oh I don't know what I'll do if I can't read anymore."

I have a good personal example of the non-collegiate learner. Less than a year ago I was called by one of my best boyhood friends, a fellow I have not seen for fifty years. He had heard about me because of a book I had published. He and I have traveled totally different paths in our lifetime. He followed the training of the high school we both went to, he became a tool and die maker. And as I indicated, I went on to college and ultimately became an academic. Our lives could hardly have been more different. Even so, he also became a reader and I believe he learned to play the idea game. I now talk to him every couple of weeks, and find our conversations very pleasant. Our ideas run down parallel tracks. He may lack the background for some of the things we discuss, but he understands anything I can tell him; and I find his ideas analytically sophisticated. He thinks. So for him and his kind also I dedicate this book, even though I wrote almost all the pieces without even knowing he was alive.

There are of course a few natural seekers, or ones who have already well cultivated the art of seeking, in and out of college. To them I have little to say except thanks for being in my class or for reading my book. Some are young, as I was when I was so overwhelmed by the pleasure of learning in my first college classes; while others are old, as I now am, but still delight in the joy of learning. And we seekers can play together in the field of ideas.

A contemporary problem is the era. Although economic matters are by no means as bad as during the Great Depression, we have been living in fifty years of uncontrolled consumerism, when the Boomers and Generation X came to believe that the standard of living would continue to go up forever. Furthermore they learned that the supreme achievement of life was to achieve the highest level of consumption. There is only now beginning to be heard in the land a small voice saying that using only what one has is justifiable. We now know that the gross consumerism of this era depended on using vast resources of a world which only has a limited amount. But the main message, abetted by most economists, is still to consume as much as possible. And thus one must earn more, and to get that, one should get as much specialized job training as possible. There is only a little space left in education programs to learn to improve one's future life quality. Thus college curriculums are being tailored to make them more specifically into job training programs for employment. Colleges still have general education courses, but I think most administrators view these as little more than providing background for employment. I have recently learned that the two-year colleges are raising particularly high the fees for returning middle-age students. Presumably this is because these life learners are less concerned with occupational goals. Anyway, I have been out of step before, so another time will not be that important. And I am also in the fortunate position that I do not have to shore up my career by writing works that would assist me in getting a job. But even so,

Introduction

I must emphasize that I became a player in the field of ideas rather than training for employment very early.

So, this piece is a plea for learning for life, for seeking out and cultivating ideas, for being a fully human animal. And the essays which follow will attempt to show how one person engaged in such a journey during a lifetime. It might serve as a model, but with the understanding that ideas and learning can come from any source. It is based on anthropology only because that study of humankind has been particularly fruitful for me. And the final message is that if we consign learning, formal and informal, into being no more than a means of achieving a higher standard of living, we will end up with a culture of mindless persons who earn large salaries. We will be abandoning our human patrimony.

1

Concepts to Live By

As I described in the introduction, humans cannot do nothing for long periods of time -- at least when they are awake. When the Kentuckians whittled the house down, neither I nor the others on the street could understand why they were doing it. It was only later, after my mind was filled with many new concepts, that I understood how human they were being.

And so it goes through existence. We see and understand things according to the concepts we then have. The concepts are what make the experiences and observations meaningful. What was invariably missing were new concepts to channel the perceptions. One of my early friends, a thoughtful person in his own field, mechanics, used to categorize American Indians negatively, claiming that they were a particularly lazy people. They were not productive workers like Euro -Americans. At that time, the 40's, the true workers of the world were thought to be Americans. I felt uncomfortable about my friend's idea even then, long before I had the slightest awareness of anthropology and its concepts. But once I learned of the processes of acculturation (modification of one culture by another) and deculturation (loss of one's culture), I understood much better why Indians do less than Whites. Any people whose way of life has been systematically destroyed, not to mention other disabilities forced on them, will tend to become apathetic. They do not make good workers. The idea of deculturation made sense to me in understanding the characteristics of the American Indians. My friend never got that concept.

My sister likes to travel. She goes forth as a standard American tourist and writes me of some of the things she has seen. She is curious but lacks many concepts to channel her new observations. She wrote me recently after a visit to Peru of her amazement at the wide variety of potatoes she saw there. She obviously did not know that the origin of all potatoes, including the ancestors of the Irish, Maine and Idaho tubers was in the highlands of South America and a part of the great domestication of plants in the New World Neolithic, one of the greatest achievements in culture history. This was a concept I got very early in anthropology and into which I have been plugging observations ever since.

So we humans filter our observations through concepts. One philosophical position is that there is no natural world apart from how we think about it. And for our species this is through concepts based on symbols. We are the conceptual animal.

And where do we get these concepts? In the beginning we go through a process called primary enculturation. As infants, we undoubtedly see a multi-image universe, in color, form and movement. And we probably make some unguided generalizations about it. But not for long. Very soon, each time we begin to cogitate about whatever image is before us, another image intervenes, a linear shape with a small globe on top which is punctuated by two shiny orbs, some appendages, and a pair of flexible strips of muscle which form many shapes, emitting at the same time special vibrations. We are looking at a human face, probably our mother's. And it is talking to us, saying "Baby wants bottle," or "Baby sees Mama," or "Baby is cold," or whatever. And we begin then the long human road of learning our own language which embodies all the popular concepts of our culture. That's right, the simplest concepts we deal with all the time are words. So building up a vocabulary is building the conceptual base.

By the time we reach our early teens we will be well indoctrinated into our culture, though some primary enculturation may continue until we get so set in our ways in later years that no new concepts will be accepted. We will have become Archie Bunkers. Often by then we are approaching senility. I think of this process as "brain-washing" and Mom as the chief brainwasher. What she did to each of us, with all the best intentions, makes the Russian, Chinese or North Korean intelligence officer look like a rank amateur. She of course had some advantages, primarily that we had not already been "brainwashed" and also that she could concentrate most of her efforts on the one of us. More or less at this stage we graduate beyond words as concepts, to statements. The building blocks are still words, but they are put into order to make up generalizations. Thus we learn that "The world is round." Not flat, as in a previous concept. And that "Human beings are the only talking bipeds on earth." We are also introduced to the human concept of quantity, to numbers, which can also be made into generalizations, such as $E=MC^2$. Anyway, that is how we got our first concepts, what a dog is in contrast to a cat, what cleanliness is in contrast to dirtiness, and how Moms differ from Dads, etc.

But that is only a beginning. Those of us who came into contact with peoples of other cultures may have got some of their concepts. The world's people of color have been exposed to a very outspoken, self-righteous kind of religious practitioner, the Christian missionary. And this person has transferred many new concepts which have been at least partially assimilated: that there is only one god, that he is three gods in one, that only humans have an indestructible inner substance called a soul, that the human body, and particularly that of the female, is fraught with evil and temptation, etc. A different set of concepts regarding government and human rights have been transferred by politicos and teachers. Our parents continued to give us concepts also. The list goes on. We white persons even have a few concepts from the people of color which our ancestors subdued, but of course far fewer than they from us.

Concepts come with specialization of occupation, particularly in complex societies. They are part of the whole package which one gets on learning the job. If one becomes a psychoanalyst, one learns that the subconscious is a powerful driving force, if one becomes a Fundamentalist minister, one learns that the Bible is the word of God. If one becomes a member of the hard sciences (chemist, physicist) one learns that there is only relative truth, based on observation, and if one becomes a doctor of Western medicine, one learns to keep individuals alive as the highest good, etc.

The concepts we get serve both as blinders and points of focus. On one hand, they frequently prevent us from seeing in other ways. If we have a concept that the world is flat, we are unlikely to make a voyage west to get to the east. Thus, one of the main new concepts that changed our world was that the world was round.

Creationism is a concept that inhibited new ideas for over 1,000 years. The idea that the universe was just like it was when it was created, and that the world was the center of the universe, held back modern concepts of astronomy, biology, and culture for several hundred years. If there had not been a shift to the concept of evolutionism, along with reliance on observation for basic information, the human animal would not have got to the moon nor would he have developed the technique of gene splicing. Change in concepts can open totally new vistas. Just as some can serve as blinders, others can serve as guides. It is the positive function of concepts that I see now after more than 40 years of exposure to anthropology.

I am well aware of the many limitations to this way of seeing man's place in the world, but this body of concepts has opened vistas for me that I could imagine getting nowhere else. I am sure there are practitioners of other ways of knowing who feel convinced that their fields have also been eye-openers. No argument. All I can claim is that anthropology has been remarkably satisfactory to me, and I venture to guess for the great majority of others who have followed its road. Also I can hardly claim that my view of the vistas it has opened for me is the same for all other practitioners. But I do think that I am pretty much in the middle of the road, and where I deviate I will try to point it out.

The concepts of anthropology are by no means the only ones which have influenced me. As always happens, I went through basic conditioning, in my case as a Midwestern boy of the working class. Perhaps the most specific aspect of this brain washing was Catholicism. My devout parents tried in every way they knew to make me into a good Catholic. There are a multitude of theological concepts that make up the philosophical framework of Catholicism, but to me the over-riding ones were that basic knowledge had already been revealed and that the whole structure of ideas and practices of the belief system had to be accepted on faith. I rebelled and rejected these two concepts. Apart from the Catholic background, I acquired most of the standard American ideas in my growing up, that a "democratic" republic was the best form of government, that individuals had inalienable rights, that personal happiness would most likely be achieved through a romantic attachment to a member of the opposite sex, etc. The only remotely cross-cultural experience I had before being exposed to anthropology was as a soldier in World War II. It made me aware of some other human ways, even though now the differences between American and European society seem hardly earth-shaking. However, I got no useful concepts from the military experience that I am aware of. Whatever true understanding of the military and warfare I got came after intense exposure to anthropology.

The curtain parted with the first course and first book of the study of man. And through the years the conceptual base has become broader and broader so that now I find it impossible to think non-anthropologically.

What is this great body of concepts? I suppose the best way is to present them more or less in the order I got them, as the anthropological lens started to come into focus.

First, I learned that the field of study was exclusively concerned with humankind. I had known so little about it before that I really had few preconceptions. Perhaps I had confused some part with geology and biology. Both connections are fairly natural since there is a part of anthropology that gets its data out of the ground, archeology, and the early presentation of the great Charles Darwin made clear anthropology's indebtedness to biology. Also, I very early understood that anthropology was concerned with humankind through all time, long before written history began. And indeed through the years I learned ever more definitely that anthropology was the one study of humankind that knew no time or space limits. From the era when a small, hairy primate started to raise itself upright and shamble across the East African savannah, until the present, all human eras were grist for the anthropological mill. And any place where such events occurred were okay so long as they were on the planet. But now even that bridge has been crossed. There is an extra-terrestrial anthropology.

Perhaps at first I still had one common preconception, that

anthropology was especially concerned with primitive people. And it has been historically true that anthropology has concerned itself with technologically simpler peoples more than other types of mankind. In fact, anthropology was once called "the science of leftovers." That meant that sociologists, political scientists, economists, historians, etc., concerned themselves with the customs of civilized man while anthropology, being Johnny-come-lately, took what was left, the tribals. Though no longer true, this did mark the field.

One particular characteristic that anthropology got from studying the preliterate peoples, the current more preferable term, was a concern with the broadest view of mankind. One did not throw out a custom because it was not practiced by "civilized" people. By limiting one's study to urban peoples, one reduces greatly the variety of behavior possible for solving basic human problems. Take for instance eating habits. Though quite a few preliterate, and even a few urban cultures, practiced cannibalism, the great civilizations of Europe did not. Thus, without looking into the practices of the simpler societies, how could an extraterrestrial ethnographer figure out what function this custom had? By ignoring the custom on the grounds that it was uncivilized, a not unusual practice, one would get into the bind of ethnocentrism about which there will be more words later.

Another kind of custom which turned out to have an enormous range of variability was kinship systems. Anthropologists quickly learned that they would get nowhere if they referred to family members as aunts, uncles, cousins, nephews and nieces. One couldn't understand the variety of kinship roles and terms by basing descriptions of other systems on the Western model. Thus a whole separate way to describe the chain of relatives evolved. The same with language. Anthropologists learned that apart from the languages of Western civilization, that there was such a variety, especially among the preliterates, that using Latin or English as a models was totally nonproductive. A whole new way of language description evolved. So anthropology became the social science committed to the broadest range of interest in time, place and complexity. Any creature that could be remotely considered human, living or dead, at any technological level, and in any place, was grist for analysis.

The basic concept that underlay all this variety in time and space is evolutionism. All the term really means is change through natural processes. However, many other meanings have been attached, including progressivism and gradualism. There have been differences of opinion about the secondary meanings but very few practicing scientists will argue against the basic idea that all life forms are the products of natural change processes.

This may sound like a very obvious idea to moderns but it has not always been this way. Most of the cultures of the world, including the preliterate ones, have managed with variations of the concept of creationism. That is, the world and the life forms in it were created by some powerful force in the past and have continued unchanged or degenerated since then. But, it is worthwhile remembering here that there have been multitudinous widespread beliefs that have been proved patently wrong. These include the beliefs that the world was flat, that it was the center of the universe, that malaria came from bad swamp air, and that life is frequently spontaneously generated. Most Americans have heard little apart from the creationism of Genesis. But most of the preliterates as well as other urban cultures have had variations of the same, including Hinduism and Shintoism. And of course the creationism of Christianity, as well as Islam, derives from Judaism, the recently literate culture of the Old Testament super-tribals.

The true preliterates had a wide variety of creation myths, as they did of so many other beliefs and customs. The Australian aboriginals had a poetic name for the era of creation, Dream-Time. Then life on earth was as in dreams, when newly created humans and animals were like one another, living together in harmony. Of course this all went by the board when the aboriginals were converted to Christianity, the religion of the conquerors. Then the aboriginals had to accept the "dream-time" of the White missionaries as described in the Bible.

I have always found one creation myth of the American Indians particularly evocative. It has been named by folklorists as the Legend of the Earth Diver and points out the feeling of kinship between the Indians and the animals among whom they lived. It goes as follows: The earth was covered with water and uninhabited except for one creature, a kind of muskrat, which was swimming on the surface. Earth Diver got lonely and plunged down to get some earth in its paws or mouth, bringing it to the surface, depositing it, and creating an island. Down again to get a tree, bringing that up to stick into the drying earth to begin a forest, down again to get an animal, up to deposit this new life form. Thus, one after another, the things that are on earth now, including humans, were brought up and started off. It's Genesis all over except that an animal form is the creator rather than an all-powerful anthropomorphic god. Also the details vary. Anyway, this and other creation legends, most of them recorded by anthropologists before they faded from cultural memory presented the world as an unchanging or deteriorating place since the time of creation.

So, when Darwinism, the idea of continuous change without the intercession of a supernatural figure, was offered to humankind, great shock waves spread, first in the West, later elsewhere. Most of us know that Darwinism was strongly resisted by creationists and still is by Christian Fundamentalists. It was, and still is, a shocking idea, including to many who do not actively resist it. And though almost all biological and social scientists, as well as many other scientific types, accept evolutionism as the best concept to explain how we became the way we are, there is still enormous resistance to the idea.

Evolutionism is the idea that provokes most resistance in introductory anthropology classes. This is not too difficult to understand. Creationism came to these students, as it did to me, from their primary enculturation, from Mom or Dad. It was perhaps abetted later by the minister. Most of us got this idea when we had no capacity to evaluate it, as is usual for all primary cultural conditioning. It is extremely unlikely that we were told the alternatives at that time. Thus, we learned that the universe had been created by god, what was good and bad to eat (hamburgers good, earthworms bad) and that truth was good while lies were bad. Is it any wonder that many of us resisted the idea of evolutionism presented by a college professor twenty years later? We were then taking an anthropology class to fill the social science requirement and get a degree in computer science. The reader might say, "But you claim that you got the same idea through early conditioning and you say you changed."

That is true, but my case is different from that of most others. For all intents and purposes I rejected that part of my brainwashing.

For whatever reasons, the creationism of my Catholic boyhood was not satisfactory. Furthermore, I went to college to seek new ideas, not to get a higher level job credential. So when the professor presented a simple version of Darwinism, I was ready for a change in concepts. My attitude was wholly different from that of most of my students years later who had never doubted most of the explanations of their parents and did not come to college for basic new concepts. I know there have been some doubters in my anthropology classes. There always are, but they are invariably in the minority. But without these few, who would want to teach a "useless" subject like anthropology?

Anyway, the idea of evolutionism is considered one of the half dozen great ideas of the modern, scientific world. It practically ended the concept of fixity of life forms in the centers of intellectualism.

Charles Darwin was a biologist and drew his conclusions about living forms from the past and present. Thus, the idea explained how living groups came into existence. He clearly recognized that humankind was one such. Thus the origin of the creature called *Homo sapiens* could be explained through the concept of natural selection. And since in the early days anthropology was much more concerned with physical man than with cultural man, it was natural that the new concept be applied to the study of the hairless, talking biped. And the concept has remained the bedrock of physical anthropology.

To me personally Darwinism provided an acceptable alternative to the concept of creationism. Furthermore, the more I have learned of the physical changes in mankind through the last four million years, the more comfortable I have become with the idea of natural selection. How good of my anthropological ancestors to have pinched this idea from biology. Of course, other scientific types did also, so that today Darwinism is spread throughout intellectual circles.

But evolutionism meant more to anthropology. Humankind has always recognized change more easily in customs than in physical characteristics. After all, any anthropologist of the 18th or 19th century, with just the slightest awareness of history, knew that Roman customs were different from contemporary Italian ones, even if a modern Italian is a descendant of an ancient Roman. So it was easy enough to think that a contemporary Italian was physically the same as an ancient Roman. But from the steadily growing subfield of anthropology, archeology, as well as from historical studies, there was continually increasing evidence that civilizations rose and fell. It was very clear that cultures did not stay the same. Thinkers had long tussled with the problem. Moreover, they were hampered less by the theories of creationism than when biological change was considered. After all, the classic theories of creationism dwelt almost exclusively on the origin of physical entities. Genesis does not bother to tell us when the ancient kingdoms of Egypt and Babylonia were created, but does give a general list of physical features, plants and animals that were created.

But since then various theories of the rise and fall of civilizations have evolved. And what would be more natural than that the new concept of evolutionism be applied to civilizations. They rose and fell just as species did, by natural selection or social selection, the functionally efficient ones replacing the functionally inefficient ones. The idea became entrenched and we still have it in various forms.

And that brings me to another idea in anthropology, cultural relativity, which runs counter to evolutionism. Relativity means that occurrences exist only relative to their context. Einstein's famous $E=MC^2$ refers to the idea that time and velocity exist only relative to one another in the context of energy. Cultural relativity means that customs can be understood only relative to their context, which we generally think of as the whole cultural pattern. The idea is that one cannot claim that cannibalism is either good or bad except in reference to the other customs and goals of the society, say Aztec or Melanesian. And a system of nature worship must be as good for some people as a system of anthropomorphic gods for another.

This idea clearly went against evolutionism, which entailed the belief that more efficient forms replaced less efficient ones. Placental animals (tigers, deer, bears, rats) replaced marsupials (koalas, kangaroos, wombats, wallabies) in most of the world. But relativity went counter to another idea also, one which anthropology has been much concerned with, ethnocentrism. This concept, which is still very much existent, is that individuals tend to evaluate other peoples' customs in reference to their own, and usually negatively. That is, customs from other cultures, if different, are judged not as good as one's own. This idea is obviously still widespread in the cultures of the world despite some verbals admonitions against it. Christians see nature worship as inferior to their own anthropomorphism, non-cannibals look down on cannibals and stop the practice when they get in control, vegetarians judge meat eaters as less good than veggie eaters, and certainly people from constitutional democracies look down on Marxists. Furthermore, each group has provided its followers with "proofs" to justify their position. Ask any Hari Krishnaite why one should be a vegetarian. But of course if you ask a meat eater, such as an average American, he too has a "reason" why it is good to eat meat. Furthermore, any good Christian can "prove" why his creed is better than sun worship.

Some might say that the decline in the justification of Marxism in the Soviet Union in recent times belies this point. I don't think so. It simply shows that "proofs" change when conditions change. After all, Europe got by with monarchies for many centuries, along with the appropriate "proofs." But after social upheaval and change to elected governments, a new set of "proofs" were generated.

And without going into the problem of whether some customs are not truly better than others for everybody, the real fact is that in the enculturation process all peoples are given concepts of rightness and wrongness, along with explanations. These are called cultural rationalizations and seem to serve the function of making people feel right about their own way of life. I understand why anthropologists have emphasized so much the ideas of cultural relativity and ethnocentrism. Anthropology got its start primarily in studying the way of life of others. That is, while historians were principally concerned with the few civilizations that had survived, and economists, political scientists, sociologists and psychologists turned their attention to modern cultures, principally in the West, anthropologists studied all the remnant cultures of the world, primarily tribal. The anthropologists' "people" were American Indians, Polynesians, Australian aboriginals, and other pre-urban peoples. And with such a variety of small tribes adapted to different microenvironments, it was inevitable that a very wide range of customs would occur. If anthropologists had stuck to absolutism, they would have had to decide which was better, civilized or tribal life. And that would really have put them in a bind since they were from civilized societies, studying tribal ones. If they had opted for civilized societies, they would have had to look down on tribal ones, thus knocking their source of livelihood. Moreover, they would then have had to compete directly with sociologists. If they had opted to praise tribal societies exclusively, they would have had to knock their own. Actually, they have done the latter to a considerable extent anyway. It has been usual for anthropologists to criticize their own culture in comparison to others, as often as not tribal societies. This is of course just the opposite of non-anthropologists, who though they may criticize their own culture, practically never do so in comparison to those of preliterate tribes. My wife has caught me up on this twist quite often.

But the anthropologists found a way out, with at least a temporary solution. They invented a fancy dance called cultural relativity. According to this doctrine, all cultures are okay, judgement of customs being proper only in terms of each society's values. This got anthropologists at least partially off the hook.

It may seem that this is merely a case of special pleading, that there is no validity in the idea for explaining culture. And while I do see that this concept served anthropology by letting them stay on the fence, I also think there is some legitimacy to it. The real fact is that non-anthropologists, and particularly nonsocial scientists, tend to go in the opposite direction. That is, they judge customs ethnocentrically, according to the values of their own societies. They may agree that ethnocentrism is not nice, but watch them when they get down to the nitty-gritty, to the customs which matter a lot to them. How many times have we heard of a Christian minister who viewed tribal totemism as just as valid as Biblical dogma or an American politician who saw Marxism as acceptable as Western democracy, or a Western dietician who viewed cannibalism, insect or dog eating as just as acceptable as cow eating? No, I'm afraid, for the basics of life, men depend primarily on ethnocentric judgements.

Ethnocentrism has entered into the vocabulary of Western humankind to a certain extent in this day of nondiscrimination. Since the end of World War II and Western colonialism, we are supposed to have moved closer toward a world where it is not cricket to discriminate on the basis of physical characteristics or customs. Perhaps we have come a certain distance along this road. Probably the best way to discredit an American politician is to accuse him of sexual dalliance or racism. But most of us also know that the average citizen has hardly turned a flip-flop in racial matters. And we also know that though Christian leaders may talk of ecumenism they have certainly not given up their efforts to convert pagans (their term, not mine). And I suspect there are mighty few Americans who accept unequal treatment of females in the Japanese marketplace solely in reference to Japanese culture. I hardly need mention what would happen to a restaurant which served ground horse meat in the U.S.

Of course we are not ethnocentric about little things. It is chic to eat some exotic food when we travel: black beans and rice in Costa Rica, cherimoya or jackfruit in the tropics or whole grain rice in a hippie commune. But let's face it, we try to get most of our food, even while traveling, in the Holiday Inn, Hilton, or McDonald's. And though we make a to-do about eating with chopsticks in a Chinese restaurant, and some of the really brave may use their fingers when trying Middle Eastern food, we know even then that we will be back soon to the good old knife and fork.

Is ethnocentrism so bad? Not really, at least so long as it does not bring societies to the point of committing mayhem on others. But of even that I am not sure. The idea that each people or society has some inalienable right to continued existence is in itself a relative value and certainly goes counter to the basic tenet of evolutionism. Societies of the past have usually tried to eliminate their weaker neighbors and certainly were ethnocentric. What Greek, Roman, Chinese or Inca considered the barbarians on their borders as having equal rights? And the powerful ones inevitably imposed their own customs on those they conquered. This process has of course continued right up to the present, no matter how often self determination was praised as a right. How many American or Russian officials conceived of permitting American Indians or Asian nomads to maintain independent cultures? Even the immigrants were pressured into a mold. The pressure to force English onto the sons of immigrants in this country was formidable. I had a friend in college who had been corrected so harshly in class for using German words, the only ones he knew at first, that he developed a permanent stutter. And no-one denies that the African slaves were forced to accept a form of English, while the main function of the American Indian education program was to compel them to speak English. The only real exceptions have been large ethnic groups, like the Hispanics in the U.S. and the French in Canada, but they maintained some of their own customs despite Anglo pressure.

In any event, ethnocentrism seems like a natural attitude, produced through primary brainwashing, and serving in the long run to help maintain the integrity of a culture. Americans rally around the Constitution. This is not because of its intrinsic value, which has been thoroughly eroded by amendments in any case, but because it is an issue to foster patriotism.

But ethnocentrism goes against the idea of cultural relativity. The pragmatic though sloppy solution is to be sometimes a relativist (totemism is just as good as Christianity) and sometimes an ethnocentrist (all peoples have basic human rights). This is what most anthropologists and liberals do, and it works most of the time, though least well when we try to apply evolutionism, and particularly the concept of natural or social selection. After all, if customs are naturally or socially selected, some must be better, that is, more fit than others. Anyway, anthropology shambles on with these conflicting ideas, much as I picture early man doing, sometimes upright but then again on all fours. As a species, he finally made up his mind. So perhaps someday anthropologists will make a choice between relativity and absolutism (ethnocentrism). It is even possible that the rest of the populace will decide also, although up to now it has hardly been a problem with them.

Another idea I got early in anthropology has also become popular. It isn't so much a new concept as a counter to an old one. Like most of my peers, I grew up in a world where people were judged by race or a combination of race and culture. The two were mixed constantly, and since anti-racism has become a popular rallying point, they still are. My mother carried along the cultural baggage of most other Americans. Thus, one could identify a Jew by his long nose, and if more proof was needed, one could fall back on behavior -- he cheated you at the cash register. She did not worry about what connection there might be between a long nose and dishonest behavior, as did few others. After all, her knowledge of the behavior of Jews, or even their identification, was not based on any deep awareness of their history. She certainly did not know that the so-called Jewish nose was just as likely to be found among Arabs and other Mediterranean peoples. Judging people by their appearance and mixing learned behavior with inborn characteristic is certainly an old trait. But it seemed to reach its zenith not too long after Euroman spread throughout the colored man's world. By the time I began to study anthropology, it was no longer cricket to attribute behavioral characteristics to inborn traits. The Jewish nose became irrelevant, as did white skin, slanted eyes and thick lips for explaining a people's cultural condition. Each people was the way it was because its members had learned how to handle money or sail ocean-going ships or organize hundreds of millions of people effectively or develop a viable pattern of village life in tropical rain forests. The accent was on learning, not genes. We had come to the era of culture.

So another new concept: the human creature is what it is because it has a culture. Although there may be a biological base, this is modified by what is taught. Furthermore, though the biological base remains the same, no matter the physical appearance, the learning varies from culture to culture. Thus, like other animals, humans have a biological drive to consume food. But whether this is steamed rice, wheat bread, boiled plantains, or corn tortillas, will depend on what is learned in primary brainwashing (enculturation). The same is true of politics, family customs, religion and art. There is thus no natural man, only a genetically controlled creature which learns the customs of its group.

Some of my readers may say that they don't eat what they learned as a child, or have the same customs as their family members. This is very possible. In fact I have changed my eating habits completely in my lifetime. And the reason is that we only start with primary brainwashing. But as we go through life, we get other concepts. Read how I became a vegetarian in a later essay of this book. But in any event, the primary, secondary, tertiary and all other concepts we get in a lifetime are products of learning . Basically what we get from the genetic code is the capacity to learn. Some people like handy definitions. So here's one. It is also a major concept. "Culture is the sum total of learned behavior shared by members of a group and passed on from generation to generation with modifications." There are of course other definitions.

Anyway, this is a central concept in anthropology and one which made a lot of human behavior explicable to me as soon as I understood it. Most of our perceptions of the universe and the behavior that follows are consequences of our culture.

This idea has made relatively little headway in Western Europe and America. It goes against the predominant value of our society, individuality. But of course individuals are members of groups. And no matter what level of biological complexity, individual organisms try to continue their existence as long as they can. Not only does the cottontail try to outrun the coyote, ordinarily with success, but the grasshopper tries to escape the bluejay and the lizard tries to get away from the roadrunner. Most get away, a few do not. Anyway, the Darwinian concept of natural selection tells us that the quick survive and reproduce while the slow become some other creature's food. Thus, new species evolve. So what begins as a drive for individual survival ends up being the basis for the emergence of new groups, what biologists call species.

Humans are no different. They too have the biological drive to survive as individuals. Thus, individualism must be thought of as a natural animal characteristic. However, humans, as well as quite a few other creatures, are also social animals. And invariably a social group exists only by suppressing some individual needs. In the same sense that the baboon group survives by keeping immature males on the outer fringes of the troop, humankind protects its territory or grabs that of others by sending its young men to war. It is certain that the immature baboon would like to have free access to females in heat and young men have no wish to be killed in warfare. Thus, individuality is in basic conflict with social reality. Evidently from the pressure of natural selection, some species simply do better as groups than as populations of individuals. Thus some individualism is always suppressed in social groups.

However, there is variation in the permitted freedom of behavior in different societies. Some permit relatively free sex before marriage while others prohibit it totally before the social ritual, some societies promote religious exclusiveness, while others are eclectic, and some emphasize group commitment while others focus on individual needs. Japan emphasizes loyalty to the group, while America concerns itself primarily with the rights of the individual. The primitive Eskimo and Yanamamo were just as individualistic as the frontiersman or acquisitive capitalist, but the ancient Inca, Chinese, Hindu and Roman owed their loyalty to the social group, whether the super-family, clan, caste or ruling class. And thus Americans resist the idea of group customs controlling their behavior. Each wishes to think that he/she takes action because of an individual choice. The standard Anglo-American feels much more comfortable explaining that he doesn't eat spicy food because it gives him a stomach upset rather than that ninety-five percent of his compatriots have the same custom. He is little interested in the fact that ninety-five percent of Mexicans eat spicy food with no apparent stomach upset when he is explaining his own behavior. I recently had dinner with an Anglo-American lady who insisted that her preference for blue eyes was purely personal. There was no way I could convince her that the usual preference in the society is for blue eyes, that the cosmetic procedure of using blue lens to cover brown eyes and not brown lens to cover blue eyes was a reflection of a cultural bias in the culture.

The particular concept that most behaviors are a consequence of group membership and cannot be explained as matters of individual choice Americans find difficult to accept. Each maintains that others of the society may act socially but that he follows his own tune. This is not the only instance that anthropology finds it difficult to pass on its message but it is perhaps one of the most critical. But this should hardly make one despondent. Ideas that go against the grain are rarely accepted easily. If a few men had not persisted, we might still believe the earth was flat, that individual species had remained unchanged since creation, and that a creature as rational as man could not possibly have a powerful hidden self, a sub-conscious.

Along the way I also learned that culture is not a hodge-podge of customs, a thing of shreds and patches, as one of our early spokesmen claimed. Professor Wissler was interested in how cultures change. In his day anthropologists were concerned with the problem of how customs were passed from people to people. As usual, someone came up with a word for it, diffusion. So once a folktale was thought up, how did it pass from tribe to tribe? Or how did a method of pottery-making, like the coiling technique, spread so widely, or how did a custom like tobacco smoking travel around the world in one hundred years.

And while others agreed that customs did indeed diffuse widely, that this was not haphazard. When not forced on them, some new ways of doing things were not accepted. One of my early professors liked to point out that the Japanese emphasis on strong hereditary social rank prevented them from borrowing certain characteristics of Chinese society. They did not take over the Chinese civil service system, though they accepted Buddhism. And in more recent times, though the Japanese were quite willing to accept and utilize Western industry, they have been much more reluctant to take over Western social practices. Thus Japan is still primarily a male-dominated society and dissatisfied workers who seek jobs elsewhere are not in favor in the workplace. From another culture, the English tradition of bland, boiled, roasted or fried foods, which was carried to America, has effectively blocked the borrowing by Americans of hot spicy sauces from Mexico, India, and Africa. All successful restaurateurs in Gringo-land know that the first rule of operation is to keep most of the chile out of the food.

Anyway, the consequence of borrowing some things and not others, and then modifying what has been borrowed, caused cultures to become systems of inter-related parts, not hodge-podges of shreds and patches. Thus, the plants the Anglo-Americans borrowed from the Indians got well integrated into the whole system through time. Take tobacco, for instance. From the time a well-meaning Englishman sloshed water on the smoking Sir Walter until the era of condemnation of tobacco as harmful to the health, the weed became so integrated into American life that a condemnation by the surgeon general has been far from enough to get rid of it. It had become sex-linked, a sign of sophistication, cigars, pipe and chewing tobacco male, cigarettes mostly female. Growers were given subsidies for cultivating tobacco and it became one of the chief taxable items for the government and a mainstay in the advertising industry. It became involved in our health in many ways, from being a chief cause of household fires to a cause of lung cancer. And ultimately it became a part of the insurance industry. Sir Walter would hardly believe it. And the upshot is that if tobacco were eliminated overnight, there would necessarily be major adjustments in many aspects of life. Much the same is true of all kinds of borrowing. The Western world, and other cultures later, did not just get a simple theological doctrine in

Christianity. It also got an elaborate system of sexual behavior, rules on how families should be run, who to marry, what parts of the body to expose, what clothes to wear, how to handle money, what not to eat, and how to run a government.

When anthropologists became aware that customs were parts of systems and inter-related, they also learned that people in other fields of study were making the same discovery. It was called the systems approach, and one of the most vociferous of the new disciplines was that called ecology. That is, an environment was made up of inter-related parts, inanimate, animate and cultural. If one part was changed, it was likely to affect many others. By ignoring this fact, industrial man got into the pollution mess.

About the time I learned that customs are inter-related, I also learned that they had functions. The functionalists claimed that no customs existed in a cultural limbo. Instead, each contributed to the ongoing system of which it was a part. Thus, a shirt and tie, covered by a lab coat with a stethoscope projecting from the pocket, helped establish the person's identity. He was an M.D., while the identity of a male nurse would be marked by surgical blues. A banker's three-piece suit, striped tie and white shirt indicated his stability, he hoped. On the other hand, a social science professor's open-collared shirt with corduroy pants or blue jeans indicated his free ranging mentality. All were markers of social roles. One's clothing style served a function. Among other things, it made individuals predictable, and helped keep each social system functioning, at least for a while. And so also with other customs.

Sometimes in order to avoid boredom and letting my lecture go stale, I would play a little game in class. I would pretend to search for some customs whose functions were not obvious. I did learn that there are only a few of these for which by going back one cannot find some function. This is despite the fact that cultural systems are not so automatic that customs are dropped the moment they lose their original function. Thus, many of us do not know the reason why we say "God Bless You" when someone sneezes. It derives from the function of blocking the entry of an evil spirit when we have momentarily lost our life spirit through expulsion of our body air. Going back to the origin reveals the function. Many anthropologists believe that people in a society frequently do not know all the functions of customs in their own culture, nor do they need to. In order to keep a culture going it is only necessary that a majority follow the customs. They do not need to know why they do everything. Rationalizations will do. Thus, the explanation that one does what one does because of personal choice is perfectly alright so long as a significant majority follows the custom.

So what does this elaborate integrated system consist of? Anthropology, like some other fields, has adopted the buzz word, holistic. That is, many of the practitioners claim that since all parts are interrelated, one can logically view only the whole structure. But this is only partially true. Logically, if all parts were equally interrelated, one should study the whole system. But they aren't. All parts are neither equally interrelated nor of equal importance. A culture's technology is crucial. For survival in the modern world, a culture has to be industrial. And though many artists may disagree, the art style of a culture is secondary. A culture's technology and social structure are critical, while the kind of music and the style of art do not directly affect a peoples' economic and military competitiveness. Whether we like it or not, these last two aspects of a society spell the difference between continuity and cessation when push comes to shove.

This is even more apparent with lesser customs. What difference does it make for a culture's survival whether the people eat with chopsticks, forks and knives, or fingers, or whether a single anthropomorphic god or a variety of nature spirits are propitiated? This is no different than with any kind of system. Though we are admonished nowadays to look at all aspects of our environment to find the cause of pollution, it is obvious that our concern with auto exhaust emissions is of a totally different order of importance than the numbers of eggs condors lay. So also, when we look at an automobile as a system, it is obvious that the horsepower generated is of a totally different order from the color of the paint job. So in practical terms we do distinguish between parts of a system, simply because they are not equally integrated or important.

Another problem with viewing systems holistically is that human comprehension, like that of all creatures, is strictly limited. Just as one cannot view the entire universe in one fell swoop, one cannot "see" an automobile or a culture in entirety. So we have evolved the procedure of subdividing, or breaking a system into components. Before the idea of holism, we hardly did anything else in viewing the universe. Now we try to "see" both components and whole systems. One time when we look at the universe, we concentrate on the moons of Saturn, another on the western quadrant of the Andromeda Nebula, all the time trying to keep in mind how the one fits into the whole. With an auto once it might be the ignition system, the next time the brakes. And so too with cultures. Despite the advantages of looking at such systems holistically, we end up practically by subdividing them.

So, we have sub-systems in a culture which can be further subdivided, and again, and again, until theoretically we get down to the individual components, the customs. What are these sub-systems? First, there is language. This is one in which Biblical writers and social scientists are in agreement. We know that all peoples of the world when contacted after 1492 had fully formed spoken languages, though only a few had writing systems. It is almost impossible to imagine a culture functioning without such a complex communication system. So we feel that symbolic language underlies all the other customs of a culture. Second, there is technology. Basically this is the tool-making tradition of humankind, evolved to exploit the environment. It is critical. Without it, we would starve, freeze or broil in most climates. Third, there is a culture's social organization, the customs that enable its people to live together in cooperative groups. These three are the primaries. There are also parts of culture we think of as important, though when it comes down to the survival of the group, they can easily be replaced or ignored. Already mentioned in this regard are art traditions. And though I know that many believers will disagree, another such cultural component is religion, or what we anthropologists call the supernatural. Mormons, Shiite Muslims and Hindus may equally believe that their particular faith is the critical center of their culture. But their survival as independents has and will depend much more on their technological capability and the effectiveness of their social systems. Their religion may have given them the drive to conquer vast areas of the world, but their ability to do so depended on their technology and ability to organize men. Christianity and Islam make two excellent examples. The Spanish conquistador and Arab horseman may have been driven by their religious fervor, but their ability to conquer was a product of their fighting ability. The customs of dominant cultures get spread widely, no matter what. The one great tradition that has had the most impact in the last 500 years has been Western. Thus, we find Western art, music, religion, literary styles, and language all over, not because they are better but because they were carried forth by Euroman. This was a consequence of Western technological and social dominance the last four hundred years. The places where Christianity got established were "opened" by conquistadors, navies, cavalry and other forms of soldiery. The advocates of "the prince of peace," the clerics, normally got established or managed to stick it out by hanging onto the coattails of their armed colleagues.

There are still other aspects of culture which are of secondary or even tertiary significance when it comes to cultural competition, customs of play, games, entertainment, fashions, body decoration, and more. These are more or less integrated with the technology and social organization of their respective cultures. Probably a general rule for understanding a culture is to look first at its technological capability, then its social system, and last the other components. Perhaps a good question to ask is "What keeps all these parts sticking together?"

What integrates them? Other systems, such as autos, are generally integrated. The sources of the machine's integration are the power system and electrical circuitry. Since a culture is a system of ideas, it is concepts, even if grand ones, which hold the works together.

The super-concepts are called values. These general beliefs guide conduct and help integrate the culture. The members of the culture will not always be aware of their overall influence, though some are readily apparent. It is difficult to imagine a member of an Islamic society who does not recognize the overall importance of the "will of Allah." Also, it is difficult to imagine the American who does not know of the importance of individual rights, or individualism. Hank Reardon and Henry David Thoreau were as American as hamburgers. Anyway, such values seem to exist on all societal levels of all cultures and provide guides for conduct. Because of our value on individualism, we Americans practice what social scientists call horizontal mobility in our work lives. Thus, we move to another job if we are not satisfied with the current one. The Japanese look with great suspicion on such behavior, since it reflects lack of commitment to the group. The hero is a "company man," not a rugged individualist, one who carves out his own empire through his own efforts. Thus they practice vertical mobility, moving up (and down) in the same organization. This satisfies their value to group commitment.

Thus, while fulfillment of individual potential satisfies Westerners, fulfillment of familial responsibility satisfies Middle Easterners and South Asians. When there were still viable American Indian cultures, those on the Plains, tribes such as Blackfoot, Kiowa, and Sioux, focused on aggressiveness among males as a primary value. Like everything else, values change. Thus, in North America the value of sharing consideration has replaced the older one of male dominance in the family system. The laws have changed to reflect this social change, as has literature. The sensitive male lover is now popular in modern romances and the soaps, replacing the no-nonsense, takeover male. The bride no longer "obeys." Instead, she "cherishes," as does her new spouse. The Latin male who beat his wife when drunk, or the Comanche Indian who cut off his errant wife's nose, would be given short shrift in a U.S. court, no matter how loud the proclamations against ethnocentrism.

All members of a society do not always live up to the values. In fact, one of the chief causes for social unrest is probably the discrepancy between ideals (values) and real conditions. Although it is widely proclaimed that deprivation is a primary cause for crime, a very plausible alternative explanation is the discrepancy between what people are promised and what they actually get. In the U.S. there is the widely proclaimed value of equal rights and opportunities, irrespective of race, religion, sex, etc. And although there are ethnic groups which certainly have less because of discrimination, by world standards the lower classes in the U.S. are well off.

We know that the ghetto riots of the 60's consisted to a considerable extent of poor blacks carrying off luxury items from non-black stores. The rioters were not hungry. They were more victims of a conflict between the real and the ideal. They had been told all their lives, primarily by the media, that they had equality of opportunity. But obviously they did not have equality of possession of electronic devices. After the slums I had seen in the eastern U.S., not to mention the shanty towns overseas, I had difficulty seeing Watts as a ghetto. It was certainly not Beverly Hills, but the little box houses, most surrounded by grassy lawns, with big cars parked in front, were a far cry from the dilapidated tenement slums of Chicago or New York, not to mention the shanties of workers in India where I did my first field work. It is significant that though the amenities of life among workers in India was far less than that in Watts, the Indian workers rarely rioted because of deprivation. Of course, they had never been told that they had equal rights or opportunities. And so they accepted a much higher degree of inequality.

Recently I made a visit to the old neighborhood of the 1930's in Indianapolis, Indiana. My childhood home, and the others on the street, seemed as unprepossessing as any in Watts. Furthermore, that was during the Depression when one was lucky to have a job for $15 per week. And if one did not like it, one paid for one's own retraining, as described before. I recall little talk in the media (radio) about equal rights in those days. There were no riots or looting. The values were different.

So, values like equality, individuality, moral righteousness, machismo, "the will of Allah," commitment to the social group, etc. are real. They stimulate people to action and serve as uniting forces of a culture.

The reader may say that I have just given an instance of a value serving as a divisive force. But I also said that values are not always lived up to. And when they are not, they can cause real conflict. So despite the inequalities, the value of equality of opportunity has been lived up to more often in the U.S. than in many urban societies. In general, it has been a positive, uniting force. Whether it will continue to be so in the future is another question.

There are some other less important concepts in anthropology, but no matter how interesting the topic, a writer has to stop somewhere. And even a field like anthropology can get tiresome. There were times when after worrying for days about the eight-class Australian aboriginal clan system, or the mathematical intricacies of the genetic code, that I wondered whether I had not got into the wrong profession. Fortunately, there always have been more simple, fascinating bits than tiresomely intricate ones. After the doldrums of the Australian kinship system or the mathematics of genetics, I would learn that in many parts of the world people eat clay; or the reason there are four buttons on a Western suit coat sleeve is that some smart-ass European officer of an earlier era figured it was a good way to keep peasant conscripts from wiping their noses at that place; or that one of the main problems of American baseball players bought by Japanese teams is how to get them to give up their individualistic drive to be a star in order to be a functioning part of the team.

Very early in my professional years anthropology became a way of life. I could discuss practically nothing without harking back to some anthropological concept or bit of information. I could no longer see the world with parochial, midwestern eyes, as a place of us and them. A whole new conceptual framework had been put before my eyes. Instead of a world of fixity, I saw one of perpetual change. Instead of a world of simple, innate drives, I saw one where the details of most behavior were learned. Instead of a world of ethnocentrism, I saw one of relativity, etc. And above all else, I saw a world of great variety. Certainly people were all alike. But they were all different also. I exulted in the thought. Moreover, the world changed from being one of us and them, as it had been in my youth, into a world of us only, though with multitudinous ways. I believe I have learned to live without too much ethnocentrism. My way is not necessarily best. For this I can only credit anthropology. It is indeed a different reality.

On Being a Conceptual Animal

2

The Pleasure of Its Being

Through the years many students have come to my office to get advice whether to go into anthropology as an academic major. They were invariably excited by the new vistas that had been opened to them through the concepts and data of man's way, the central focus of anthropology. Usually they would have taken several courses and frequently one or more of these had been under my direction. Sometimes they would be undeclared, not having yet decided what major to take up. At other times they would have another major, frequently one which had a clear-cut occupational goal. Some were already employed and were taking college classes as a method of getting promotions. In the university where I taught during most of my career we had large nursing and police science departments. A professional in one of the Los Angeles hospitals or the police department could use his/her college degree as a means of getting into an administrative job or at least off the hospital floor or the sidewalk. And while fulfilling the requirements for the degree, he/she would have taken some courses in anthropology, usually to fulfill the social science requirement. Simply to get the occupational degree, it was necessary to take only one or two social science courses. But the person who would be facing me would usually have taken more than the required number, using the extra ones for optional credit.

One that I remember well was a young man with a family who

was working for the sheriff's department in one of the Los Angeles suburbs. I am not sure anymore whether he took his first course with me or not but he did take several upper division courses from me, as well as from several other professors. It was very obvious that he was hooked early on. He was a good family man and I suspect he struggled early what to do about his wife and child. He must have been well ensconced in his job, which, being in the civil service, promised security. It had to be a difficult decision for him, and the last he was involved in research to collect life histories from some of his Mexican ancestors. But he was also going into education, where he could use some of the cross-cultural insights he had picked up in anthropology with Latino students. He was going halfway, primarily to use his anthropology, while still maintaining a stable economic base for his wife and child.

Then there was Ellen, an employed, registered nurse. I'm almost sure she took her first class with me. She usually came to class in her nurse's uniform with a beeper in her pocket. It went off many times and she would rush out of the classroom on a nursing mission. She became committed to anthropology very early and dedicated all her energy to its study when she wasn't working. She was divorced and supporting herself. The last I heard she was on the final stretch for her Ph. D. in medical anthropology.

Roy was different in that he had almost finished his regular employment stint. He was a policeman closing in on retirement, so getting a college degree would hardly help him job-wise. He had gone through a variety of police jobs and was a head jailer at the time he came to college. I don't remember clearly what his goal was in coming back to school, though I do remember that he took his first class with me. From that day on I thought of him as an anthropologist. The last information I had about him was that he had taken over as the director of a good sized museum in central California. I knew he had got his Ph.D. two or three years before. (See "On Becoming a Professor" for details of Roy's case.)

There were those, ordinarily younger, who were not yet employed and who got hooked. Nancy was a Taiwanese girl whose father had sent her to the U.S. to learn accounting and thus to support herself (On Becoming a Professor). She took her first class with me when she hardly understood English. She did well anyway and even better in the following classes as her language proficiency improved. Her problem was that her father wanted her to learn a subject with which she could support herself. However, accounting bored her, as she told me many times, and anthropology was so interesting. Anyway, in the end she compromised in majoring in special education, a field which had jobs but in which she could use her newly learned social science insights.

Maria did much the same thing, though she managed to get established in a closer related field and in academia. She was a Greek who had come from the working class and who had to support herself in the U.S. from the beginning. She too hardly knew spoken English when she started, but by the time she got her Ph. D., she was fluent in her new language and had in the meantime also learned Spanish, French, Italian and Ancient Greek. And though Maria too was committed to anthropology, she went on to get her Ph. D. in the related field of classics studies, primarily because of her background and skills as a Greek. Also she could get grant money for her studies in classics. She had little money for a college education. Anyway, she did well, becoming a professor in an eastern university.

When these excellent students and others had come into my office seeking information about majoring in anthropology, they would usually ask me about the job possibilities. I would try to give them a thumbnail sketch but emphasized that, for a full-blown, full-time career, they would probably need a Ph.D. And that would require 6-8 years in college. This cooled off most of them. And for those who still maintained some interest, I would say, "Of course, anthropology is not really a field where you will ever make much money for the time and expense you will put into it. If salary is your prime consideration you need to go into medicine or dentistry or engineering or some of the other practical, professional fields. So the real question is 'Does anthropology turn you on? Do you get a kick out of it? Do you think and talk and read anthropology for the sheer pleasure of it?"

We social scientists believe that in Western culture, and especially American, there is a duality in our perceptions. We see the universe morally as a place of good and bad, right and wrong, and black and white. One of the most important contrasts we learn as we grow up is work and play, the first done intentionally, highly scheduled and earnestly, with definite goals in mind, the most common being to make money. Play, on the other hand, is supposed to be done in a more relaxed manner in free time and the goal generally being only the pleasure it gives. The money from work is to enable play. We think of the work week as Monday to Friday, nine to five, the time following in the evenings and at night and the spaces in between, the weekends, as free/play time. We have a celebratory period, just after work on Friday, which we call TGIF or "thank God it's Friday." Work is over for two days and we can devote ourselves to pleasure. We aren't supposed to mix work with pleasure, to take our work home with us. And when we get finished with our working stint of 30-40 years, we can retire and devote our lives to pure pleasure. We call it the "golden years," presumably because it is the most valuable time.

I think this split in time usage is true of most Americans in most occupations. There are exceptions, of course, mainly in the sciences, theoretical and applied, and in the creative fields. For people in these fields work and pleasure are often in the same activity. And from my perspective, anthropology is near the top of the list. Work is pleasure and pleasure is work. And so back to the prospective major I say, "If anthropology really turns you on, if it boggles your mind, if you get great pleasure from it, go to it, become one."

Does this mean that a young person so inclined should not consider what anthropology will do for him as a financial career? No, I am realistic enough to know that finances will be a part of the decision. But as I said before, it cannot be the primary part since there are many professions in which the potential return will be much greater. But if the pleasure principle is clearly exemplified and there is some promise of a job, I advise students to do it. Once I got started in the first few classes back in the 40's, there was never any time I did not get pleasure from anthropology.

Now in the 90's our students are supposed to be pragmatic. They go to college to get a training for a better paying job, a profession. Of course there has been that element in higher education for some time, and especially in the U.S. This is supposed to be a pragmatic culture. If it works, we should do it.

Moreover, I have spent most of my teaching years in a city

university which was primarily dedicated to being an occupational training college. So, I have probably seen more students following the technical training idea than is true of the more prestigious liberal arts institutions. And so it is difficult for me to get across the idea that I studied anthropology all those years with hardly any consideration as to what kind of job I would get in the end. I took those classes and read those books because of the genuine pleasure I got out of it. And before some reader jumps me, I will freely admit that I had some dull classes and read some boring books, but I never did think the subject matter was dull. Anthropology was truly fascinating, even if some professors and writers couldn't get it across well.

I have watched through the years the career of a young fellow who has struggled with this dilemma since before he got his Ph. D. in anthropology. He is the same one who got an "A" in the sociology class of a hard grader. From many conversations I knew he was turned on by anthropology and other social sciences. So it came as little surprise that he went on for his Ph. D. But like the young fellow in the sheriff's department who finally opted for a teaching credential because of his need to support his family, this young "turn-on" also chose a compromise, primarily so he could take care of his family. Before finishing his doctorate, he got a master's degree in international business, as well as learning several foreign languages. There were more well-paying jobs in the market place than in teaching anthropology. So he became an executive in a foreign owned bank where he could use his language skills, as well as some cross-cultural insights. He complained several times that there was no-one in the bank with whom he could discuss anthropology.

Does the above mean that everyone who takes some classes in the subject should major in anthropology? Now obviously that will not happen, nor am I so fanatic as to think it should. But I do believe that much more has to be said in defense of studying for life, than for merely taking courses to fulfill the requirements for a job-degree. After all, how much of our lives is spent working? As I mentioned in the introduction since our culture got into the era of occupational specialization, most of the time in our lives is spent in non-work activities. About one third of half of our lives is spent in work time, without counting weekends. It isn't so much time. So how we use the non-work time is really important. Lives in which work and play are combined are no problem. We use all of it positively. Thus anyone who gets into a "turn-on" field like anthropology will look forward with excitement throughout their life to the continuous learning process. If everything about it will not be totally exciting, there will be enough interspersed to make the overall continuation worth looking forward to. The problem of interesting students, about which we hear so much nowadays, will become irrelevant. The subject will take care of that. And retirement from formal work will also be irrelevant. The great majority will want to "write up their notes," write the ethnographic novel they always put off during their teaching years, or spread the word of anthropology to the general public. Only a few will be content to put all that aside to devote their time exclusively to golf or spectator sports.

The combination of work and play, as in anthropology and some other fields, is a pattern of time usage that must have been general throughout the greatest part of man's history, as it is among the other animals. Except for a very brief period in the tribesperson's childhood, play was rarely separated from work. Thus we see the little Dani boy in New Guinea playing with butterflies while he is guarding his family's herd of pigs. On the other hand, his little sister is making play gardens in preparation for the day, not far in the future, when she will work in the real garden alongside her mother. The little Kung boy (Southwest Africa) will shoot beetles with a tiny bow and arrow in preparation for his role as a hunter, while the little Hopi girl will play with Kachina dolls to prepare herself for the day that the real Kachinas will come.

Tribal adults too combined work and play. Thus the father of the Dani boy would make toys and weave religious objects (play) while doing guard duty on the tribe's frontier (work), while the Eskimo hunter would spend long hours carving artistic designs on his harpoon. In fact, the clear-cut division of work and play has undoubtedly evolved quite recently, as division of labor and wage work became the norm in urban industrial society.

By this time there will undoubtedly be some readers who will be feeling uncomfortable because I have been harping so intensely on play and pleasure in anthropology. They might well say, "But isn't it a serious business? Is it all fun and games?"

Okay, no problem, I accept that most practitioners of the field are involved in labors to prove this or that hypothesis and to clarify further, if only a tiny bit, the enormous web of culture and its subsystems. Anthropological publications ordinarily have articles about the relevance of the subject among the sciences. It is science, even if one of the softer ones. And it can be serious business. Probably most practitioners believe that the field of anthropology has great significance for the solution of ongoing human problems. And that too is not a matter of fun and games. Again no objection, though how we are ever going to get the world-shakers to listen to us continues to escape me. But my argument is that man is a creature of multiple motivations, and one person can have several at the same time. The theoretical scientist, who is counseling world leaders, can also be getting a genuine pleasure from his field of study. Furthermore, I suspect it is the pleasure principle which brings most of us into the field in the first place. That we may become highly regarded theoreticians or advisers for world affairs comes later to some of us. But our searching graduate student is there because of his "turn-on."

So what are these manifest pleasures? To me there is little doubt that the one overall is an opening of the world. Very early in our studies we learn that there is a vast world out there of people and customs that we hardly realized existed. Our universe of mankind becomes vastly expanded. And we even get pleasure from losing some of our parochialism. Basically most cultures and individuals have operated with a parochial standard throughout history. When push came to shove, each person's and each culture's way was the best. Mighty few people have believed that alternative political systems were equally valid alternatives, much less alternative religious systems. It is popular nowadays to advocate ecumenism, or religious tolerance; but count how many believers in Christianity accept tribal nature worship or Islam as equally valid systems, much less how many Muslims see Judaism or Christianity as just as good. But in anthropology we come closer to abandoning our natural ethnocentrism, even though we do have some problems such as denigrating our own system while elevating other ones. But the overall effect of accepting other systems as valid also has been salutary and pleasurable.

We are less ethnocentric and thus find pleasure in other ways. The anthropologist tends to describe and accept other ways rather than to reject them (Is Peanut Butter Food, Love Magic). They also tend to pick up customs from the cultures where they do their studies. They bring back native garments. There are always a considerable number of anthropologists at conventions who are wearing native outfits, the women in particular. Our culture fosters innovativeness in appearance for women, while men are supposed to be more regimented in this regard. I have known a number of female anthropologists who delighted in wearing native costumes when they were lecturing in anthropology classes. And despite the cultural expectation that they will be more conservative in their clothing styles, men do participate somewhat. I have a fair-sized collection of native costumes that I have worn to parties mostly. In the field we take on the "others" ways even more, partially I suspect to get into their good graces and thus get the data about their customs that we desire. But we also get a genuine kick out of it, I suspect. Napoleon Chagnon, the anthropologist who first studied the Yanomamo, appears to be thoroughly enjoying himself when he is photographed strutting his stuff in the brief costume of a native warrior.

As a group we delight in trying new foods. At every national convention that I can remember, the local arrangements committee made up a list of ethnic restaurants which was handed out, along with other flyers, in the registration packet. And each evening one would hear groups forming to decide what food specialty they would try for dinner. I have personally tried practically everything new I could get since I became an anthropologist, until in mid-career I became a vegetarian. However, that too came about primarily because of experiences as an anthropologist (From Cow Slaughter to Ahimsa).

We are the talking animal, and so it should be no surprise that we get much pleasure from conversation or other forms of talk (A Tall Tale From the West). In tribal society the chief form of entertainment had to be telling tales and just plain gossip. But even in the age of electronics, talk is way up on the list. We have all kinds of talk shows going on in radio and television. And when we get to the individualized electronic device, the telephone, talk for pleasure is undoubtedly one of the main functions. The telephone serves business of course, but in the total amount of time usage, personal gossip is probably far more important. After all, work only takes up one-sixth of our lifetime, so there is plenty of time for yakking. When

I was a boy they characterized gossip as conversation going on over the back fence, usually between two women. What I have observed of the women in my life, it now goes on over the telephone. And though I have specified women as the prime gossipers, some men do it also, generally of different subjects, I suspect.

Anyway, one thing we can talk about, is anthropology. Get two anthropologists together and they will talk up a storm about their favorite subject. Get one and a listener, and the anthropologist will take off. I suspect I am like a lot of my fellows in that ordinary personal gossip doesn't interest me much, but if I can tie it into some anthropological framework and have a listener or another anthropologist, I can get great pleasure from such talk. I know that I, and I suspect most of my colleagues, have had the most pleasurable conversations on anthropological topics. I still remember with warm pleasure the walks I used to have with Conrad Arensberg on his Maryland farm when he would spin off an endless monolog of anthropological esoterica. I also did the same to a certain extent with my son when we were on cross-country bicycle trips. I would get started on some topic out of anthropology and one thing would remind me of another. We would be trying to keep abreast of each other at the same time. Usually he was interested, but I'm sure he got exasperated at times also. I remember him saying once after I had been going on for quite a while. "Art, where in hell do you store all that crap?"

"God knows," I answered. "It is just an accumulation of this and that from all the past years. I just think about it as I'm talking."

To touch on a still lighter aspect, anthropology makes good party talk, at least if one can get the listeners beyond the "bones and stones" stereotype. Most people think of anthropology as a matter of digging up bones or artifacts, sort of Indiana Jones without the derring-do. And though archeology and human paleontology are certainly a part of anthropology, the great majority of workers in the field are ethnologists, students of living cultures. And though the "bones and stones" people can produce some lively topics, the ethnologists can do even better. Where else can you get a "scientific" discussion of the incest taboo, cannibalism, clay eating, and the eight-class Australian kinship system? Just joking, since though real enough, that last one was mind boggling when I first heard about it.

My professor, Charles Wagley, got so confused in a lecture on this system that he couldn't figure who could marry who. He finally said in puzzlement, "But the aboriginals can figure it out," the puzzlement presumably because they were naked savages while he was an august professor of social science.

Anyway, one can spin off some good ones after having dabbled in anthropology a little while. In my earlier years I noticed it wasn't a bad method for attracting the attention of the opposite sex. After all, one of the things that American girls avoid like the plague is guys who are dull. And whatever else anthropology is, it isn't dull, even though some of the practitioners might be. So there we are, anthropology produces one of the best talk shows around.

In class I used to emphasize also that anthropology would provide all kinds of insights for choosing books, films, and TV shows. How could it not, being the mind expander it is. One is no longer locked into a parochial, ethnocentric universe. One not only looks at native or non-Western ways of life with interest, one even looks at historical events of his own civilization in a new light.

A few years ago a very good film, "The Last Emperor," came out. I went with my wife, who was very interested in personal interactions. She found the trials and tribulations of the emperor as interesting as any soap opera, a form of TV she watched whenever she had a chance. I was greatly interested in the film also, primarily because it depicted relatively well the final days of the Manchu Empire, up to the Marxist transformation following WWII. Although I am not a Far Eastern specialist, I did thoroughly enjoy an anthropology class I had at Columbia University, "The Inner Asian Frontiers of China" by the late Morton Fried. The course covered the Manchu, as well as the previous nomadic peoples of Central Asia.

Anyway my wife had become so taken in by the story that I offered some cultural/historical background to the particular events. She had never heard of the Manchus, had practically no idea of the significance of Central and Northern Asia to China as well as to Europe and South Asia, or the Cultural Revolution in Marxist China. I quickly found that I had a listener and offered to take her to a nearby place where we could have a glass of wine while I would bring her up to date on the cultural background of the film.

Later I thought that it was probable that few members of the audience knew what was going on historically in the film, as in most other films of events outside Europe and America. The film "Lawrence of Arabia," about European manipulation of the Islamic world after World War I, was probably little understood by American audiences. And the way the Middle East was chopped up then still has a direct bearing on modern events. Without the Anglo-French subdivision of the Islamic world there would have been no Saddam Hussein or a nation called Iraq.

Anyway, we in anthropology, and some other fields, have been favored by fate in our understanding of other places and times, and I think we are in a very good position for selecting books, movies, and TV shows, especially those of other cultures.

I mentioned earlier that researchers in the sciences as well as creative artists undoubtedly get great pleasure from their fields also. Probably there is much satisfaction to the individual in achievement of any kind, including those fields in which the prime motivation is greed -- money manipulation, real estate development and even criminal activities well executed, no matter the destruction they may cause. I would venture though that none would be more satisfying than the social sciences, and particularly anthropology.

And so I will close this piece with a statement of homage to an anthropologist of the previous generation, to whom I must give most credit for ever getting me into this field, Ruth Benedict. She is not so popular nowadays, and I do not disagree with some of the criticisms that were made of her work in the last few decades. But I still feel that she, like Margaret Mead, who also I suspect skewed her data, has a special place in the field and at least without whom I probably would not have got into it. When I was an undergraduate, and had just got into anthropology, one of my professors recommended Benedict's Patterns of Culture. Knowing little about the peoples described in that book and even less about anthropological theory, I was still captivated by her vivid descriptions. I thought those Dionysian Plains Indians and Apollonian Pueblos were fascinating. She was a very good writer. I did not then know that she had been a poet before. But to me anthropology came alive in a way no other academic specialty had before. I couldn't wait to learn more and even to study under this great person. I was accepted by the graduate department of Columbia University where she was on the faculty. But sadly, she died before

I got there. And so now, for whatever it is worth, I offer my homage to Ruth Benedict. She would have known well what I meant when I would write an essay on "The Pleasure of Anthropology's Being."

3

On Becoming a Professor

I frequently start a new topic by looking up the standard definition of the key word. It isn't that I swear by dictionaries, but I find them a good place to start. For this one I found definition #2, "one who professes," much more interesting than #1, "a teacher of the highest rank in an institution of higher learning."

The first sounds snooty, especially in America where a college professor is usually given a prestige rank somewhere between that of a plumber and a politician. But that's okay, no sour grapes from me. I didn't enter the field for prestige, and in retrospect I wouldn't have had it any other way. But the verb, "to profess" is interesting. It means "to affirm, or proclaim openly," especially some religious belief. That's okay too though. A lot of professors appear to be declaring some truth religiously. Probably we anthropologists are as guilty as any, even though most of us would vehemently deny that we were religious. Anyway, one thing is sure, and that is that most of us are quite convinced that we have a good corner on the truth. Furthermore, most of us are quite dedicated to profess it.

I have been as guilty as any. Once my wife and I were chit-chatting with an insurance lady whom we had just learned was an alumnus of the school where I taught for twenty years, California State University, Los Angeles. The lady was a nurse, helping her husband in insurance work on the side, and had taken a required anthropology course at Cal State. She said to Jessica, "I suppose you have taken a lot of anthropology courses."

Whereupon Jessica, who never took a course formally, said, "Oh yes, every morning." That was an exaggeration, though as usual with Jessica's observations, it did have a kernel of truth in it. I did beat her ears with anthropological tidbits, though I think I did less as the years passed. This may have been a contribution to our divorce.

Anyway, through most of my career, I actually got paid for disseminating anthropological tidbits as a professor. In retrospect, this amazes me.

So how did all this come about? As described before, I started out by making a living as best I could in Indianapolis. I had no special training, so worked as a laborer in a muffler factory, a steam cleaner in a slaughterhouse, and a clerk in a billing office. Fortunately, I got through college after that, as well as doing a military stint, and by means of some work and quite a bit of skullduggery, I got a Ph.D. in anthropology (See "Dealing at the Convention). Then I was lucky enough to get a professional position as a museum curator.

I started addressing audiences at the museum and it scared the bejesus out of me. The absolutely most dreaded class I had in my college days had been speech. That fear was still with me when I had to perform in front of the Wisconsin Archeological Society, my first speech. It was given at my chief's insistence and was on Carbon-14, then the newest archeological dating technique. I remember that I researched the topic in excruciating detail and tried to memorize everything. After the talk I couldn't remember anything I said. I couldn't believe it when the audience clapped. Later when I asked my chief if I had said anything intelligible, he said, "It came out fine, well organized and delivered." Later on, I learned to relax on the podium and cut way down on the amount of research and memorization. If anything, I got too comfortable about lecturing. But that's getting ahead of the story.

After my museum stint, I became a researcher and part-time college teacher, and ultimately a full-time teacher. My public speaking became continuous, at long stretches, and to an audience committed to learning, at least enough to pass exams. I steadily cultivated the ability to entertain, even while trying to present a core of ideas. I'm afraid I succeeded only partially, however. Although I do not think I was ever a dull lecturer, I'm sure that I wandered about more than was best, both because of my desire to keep students interested and because I just couldn't resist inserting asides from the rich repository of anthropology. It has always baffled me that any anthropologist could make the field dull, though I've heard of that not infrequently.

I worked with students both inside and outside of class. And to me this was the best part of being a professor. There were perhaps three kinds. The majority were filling the academic slot for social science with an "easy course." When I say "easy," I mean that there were not a lot of mathematical formulas, computations or dates to learn. Most students already had academic goals, in computer science, nursing, engineering, etc. They may have got a few new ideas from my class, but basically they left intellectually the same as they came in. Then there were a very small number who were already committed to the field, the majors.

But perhaps the ones I remember best are the two to three dozen who came into an introductory class of mine, either with a major in another field or still wandering around in the academic marketplace, undeclared. They got "turned on" in contemporary parlance, coming to feel the excitement I had felt way back when. They just couldn't soak in the new knowledge fast enough. Some stayed with the previously selected major, while others became aspiring anthropologists.

One of the former who I remember well was a tall, Jewish kid, doing a pre-med course. He contributed Jewish ethnographic bits regularly in class and differed with me not infrequently, but never unjustifiably. I only had him in one introductory class and I would have probably forgotten him if he had not taken a dramatic step.

I was dining in a Westwood restaurant with my wife when this young man came in with another man. I quickly recognized his face, though I couldn't remember his name. I have to know a student very well before I remember his name, though all the faces of all the students I have taught are still inscribed in my memory. He saw me also and stopped at my table. He put out his hand which I shook. Then he introduced the other man and spoke to him, while nodding toward me. "Professor Niehoff gave the most interesting class I had in college," he said.

Naturally I beamed, though I felt embarrassed also. I have

always been that way about direct complements. I said something diminishing of myself, though how emphatically I am not sure. Anyway, after a few other comments, he and his companion went back to their table. My wife and I discussed the incident briefly, then drifted to other topics. I paid no further attention to the pair until we were about to leave. When I looked to the back of the restaurant, they were gone. But when I tried to pay the bill, the cashier waved my money away, "It's already paid, sir."

"What do you mean?"

"That gentleman you spoke to paid it."

For a moment I couldn't absorb the information. People never paid for my dinner unless they were with me and then it was usually I'll pay this time, you pay the next. Furthermore, students were unlikely types to pay for professors. Finally I said, "You mean that young fellow in the back."

"Yes sir."

I never saw him again. But from what I had seen during that brief time during one anthropology class, he is now probably an M.D. and presumably carries along some cross cultural-sensitivity he got in my class.

Then there was little Nancy Hu from Taiwan, an accounting major, who I mentioned in the chapter on anthropology. She sat in the back row and very early asked if I minded if she use a dictionary during lectures. After she overcame some of her shyness, she spoke more to me. I quickly realized that she understood very little spoken English. I suggested she drop out of regular classes and take a concentrated course in English. But she hung in and came to every lecture, frantically leafing through her dictionary while I was lecturing. She got a "C" on her first exam which was fine with her and a surprise to me. After that she got "B's" and a "B" for the course. I wouldn't have believed it possible.

As Nancy became more proficient in the language and more accustomed to the informal manner of American academia, she talked more. She early on informed me that she found anthropology very interesting. She found her own major, accounting, "dull."

I asked her why she was taking accounting and found to no great surprise that it was the idea of her father, a lawyer in Taiwan. He was also paying for her schooling. His idea made sense. He was directing his not terribly attractive daughter into a profession where she would be able to support herself. In Nancy's culture, far more than in the Western world, the ideal role for females is that of wife and mother. And like it or not, the lady's appearance is clearly put in the balance.

I told Nancy, "You know your father is right. He's clearly looking after your best interests."

"I know, but it's so dull."

"But you must know that unless you are willing to spend years to get a Ph.D., you will never get a job in anthropology. Even with the degree" . . . I knew the woods were full of Ph.D. graduates in anthropology and other scholastic disciplines. She blinked at me in her myopic way and smiled, "I know."

I never believed orientals were any more inscrutable than other peoples. Most of the inscrutability we attribute to them is a consequence of our ignorance of their culture. I was sure Nancy was thinking about her problem. I knew very well that she had a strong mind of her own.

She took several more courses in anthropology, from me and from other professors, and when her language comprehension had greatly improved, she got "A's" in all courses. There had never been any doubt that she could study well.

Then she drifted away for a year or so. When we next met in the hallway, I was impressed by her appearance. She looked much more feminine, wearing stylish clothes and having a contemporary hairdo. She was with some other young ladies, and talking animatedly in English. She sidled up to me.

"How's it going?" I asked, though by appearances it was going very well.

"Oh, fine, fine." She was completely comfortable in English.

We talked about this and that, how busy she was, what classes she was taking, when she would graduate, etc. I noticed that there were no accounting classes in her list.

"Are you still in accounting?"

"Oh no, I'm a major in special education now."

That was a large department at our university, devoted to training people to teach students with special behavioral problems. It was a good field to get a job in and use some insights from the social sciences. Nancy had logically drifted into a field which was a compromise between her needs and those of her father.

"Nancy, have you told your father?"

She pinched up her eyes mischievously. "Not yet."

I kept in closer contact with the ones who became anthropologists. There was Roy, the ex-policeman and jailer who after retirement decided to go back to college, also mentioned earlier. He was older than most students, had had a lot of experience in the real world, was quite intelligent and clearly interested in anthropology. He was as much interested in the ideas of anthropology as I had ever been. He started anthropology as his major very soon after taking the introductory course with me. He then took courses from all the faculty, and so far as I know, got all "A' s". He drifted into archeology fairly early, and after that I would see him mostly in the hall, department office or at parties. No matter that Roy had already finished two careers, he was still a dynamo. Not only did he study full time, he also worked part-time as a guard in a bank and ran the anthropology club as president with an efficiency that had been lacking for over ten years. Roy also did part-time, volunteer work at the Los Angeles County Museum and was chosen to supervise the annual archeology field course by our chief archeologist. And with all these activities, he remained totally mellow. At no time did he give the appearance of being harried. And with all these accomplishments, and despite his age, Roy was selected as a student at UCLA. This is a prestige institution and it was generally believed that older students, who have a shorter career potential, as well as graduates of second-level schools like Cal State, were not usually welcome there. The faculty at UCLA could afford to be choosey. So Roy's acceptance was all the more impressive. He got his Ph.D. and went on to be a museum director.

And Maria, the Greek. And as mentioned, she too took her first anthropology class with me. And she too very early impressed me as very bright and interested. She did well in class from the beginning, despite not being totally fluent. She had just started speaking English full-time about a year earlier when she had come to the U.S. from Athens. She had one particularly troublesome characteristic. For essay exams she had so many ideas, enough non-English phrases, and such a bizarre handwriting style that I just couldn't read the answers. By that time I was already convinced she knew what she was talking and writing about so I would have her type out her answers after the exam period. She got a little better later on and in the meantime we had many a chuckle about her illegible exam papers.

I got to know Maria well during my marriage break-up and during those days I had some romantic notions, despite our age difference. Professors, like first sergeants, have certain advantages in this regard. However nothing really happened and we remained in a professor-student relationship. We kept in close contact for awhile as Maria went forward meteorically in her intense fashion. That was her only problem that I could see. Her drive was so great that I could imagine her cracking. She recognized the problem herself and we had a number of black jokes about it. However, she kept going until she got her Ph.D. in the classics and became a professor in an eastern university.

When I look at my accounts of my students, it seems that I am presenting myself as a latter day Mr. Chips. If so, it is an exaggeration. I think I have been a good teacher, though hardly outstanding. Honors were regularly bestowed on faculty members at my university for outstanding achievement. I was never nominated. But again this in not sour grapes. I know I have had certain deficiencies as a teacher, but I am also convinced that my positives outweighed them. And I do know that I have been quite content in this role.

Of course, I have been describing my success stories so far. There were negative occurrences also, enough that at times I did not want to go to class. It's like the surgeon who relates his great operations. He also undoubtedly had a number of people die on the operation table.

Perhaps the most frequent negative occurrence was midterm burn-out. Though I have almost invariably started classes with enthusiasm, sometime before half-way the course material would go flat. I would get bored with it. Usually I would have taught that course many times before and I would start thinking much of it was trivial. Then I would begin to cut. Unfortunately, the more I cut, the more I would want to cut, and eventually I would be presenting nothing fully enough to make sense. I would realize that the students were losing the thread of the topic and would think then only of getting finished.

I have always had a high propensity for that most human characteristic, boredom. This is one of the reasons I have been happy in anthropology. Whatever else it is, it is not boring. However, I have had the same problem in other activities. I like to cook but the only way I am happy is by continual experimentation. No two dishes I make are the same. Sometimes a guest will like a dish I have prepared and ask for a recipe. My reply is invariably, "I'm sorry but there is none. I never made that dish before and I don't know what's in it."

That of course is an exaggeration. If I have made a casserole of left-overs, I know whether it has peas, noodles or cheese in it. But I certainly do not know if there is one cup or two, not to mention how many spoonsful or pinches of spice. Those proportions I modify by taste. If I ever do look at a recipe, it is to learn the overall ingredients and proportions, then to modify them. Also I get bored easily with lousy TV shows and books, though a good one of either captivates me. I am very lackadaisical about reading instructions of newly purchased items. Whatever else they are, they are inevitably boring. Anyway, that characteristic can make teaching a chore and even ruin a class. I have fought it. Sometimes I have revised my lecture notes in mid-course, trying to generate a new feeling. That has worked so-so. Also I have tried to work up new examples which sometimes produced better results. But like a coward, I have even canceled class a few times to take off on sick leave. In any event, whatever I did, once I got past the mid-course doldrums and could see the harbor ahead, I would get on a new head of steam. Invariably I felt satisfied at the close of my classes.

Then there was the problem of dealing with troublesome or oddball students. The ones I described before were exceptions in both their ability and interest in the field. Most students have constituted a group I think of as the inarticulate minority. They usually came to class, took notes, and departed, contributing nothing and usually learning enough to pass the exams. Most have been majors in other fields, filling in the social science slot. They were the "bread and butter" of service courses but totally forgettable. When I would see them on the campus later I would recognize their faces but nothing else about them.

But then there have been a small number of students who

were remembered, not because of great achievements, but because they were unusual, and frequently troublesome. They became characters in professorial war stories. I have had my share.

A special type was the student who was highly specialized in some related field and who corrected errors. The ones I remember best have been in medicine (nurses usually) or biology. I crossed into those fields in medical and physical anthropology. I am not specialized in either the physiological aspects of medicine or the intricacies of biology, and particularly genetics. So, I have learned after many difficult experiences that the best way to handle such is to try to co-opt them. Thus, with a little tactical modesty, "I do not pretend to be a specialist in genetics but I believe I know more about the topic than the majority of students here. But even so, if I go too far with some of my points, I would appreciate any contribution from students who are specialists in the field."

Or, "Ms. Jones, as a nurse, you surely know the details of condition X or illness Y quite well. Perhaps you would be willing to share some of your knowledge with the class."

As often as not, the particular information offered would not be relevant. But if ignored, Ms. Jones or Mr. Smith, might still make a correction, and confrontationally. I could even have lost my place in the lecture sequence. This tactic didn't always work, but there were others which the successful professor learns along the way. This kind of thing is hardly taught in "Teaching Methods 0000," though we college professors didn't take that anyway. In any event I suspect I learned as many tactics to deal with troublesome students as most academics.

However, there were some trouble-makers so off-the-wall that only the ultimate solution was possible. That was to suggest, after an impossibly long interruption, that the disagreeing student come in to discuss the matter during office hours. Most never did. I'll relate one such. Some years ago when feminism was the rage, I had a female student who judged all my teaching, and I presume that of others, by the feminist credo. She sat in the front row and commented frequently about my "chauvinistic" statements. As I recall, I made the generalization that throughout human history males had tended to dominate females. This was nothing new, of course, it being the primary condition the feminist movement was dedicated to eliminate.

I suppose I gave quite a few examples from non-Western cultures that were new to the lady. I'm certain though that I never stated approval of that state of affairs, simply claiming that when power was concerned, men tended to take the high positions.

She would question me, "Well, what about the women? Don't they get to serve as priests also?"

I, "Generally in organized religions the controllers of ritual and other important affairs have been men."

"Well, I don't think that's fair."

I patiently, "I'm not saying it's fair or unfair. I'm merely trying to explain what is."

This kind of exchange went on day after day and frankly I got tired of it. But true to the unwritten code, I hung in, though certainly not happily.

Then one day I was explaining some puberty rites among New Guinea tribesmen that I had recently seen on documentary film. They had figured out a particularly ingenious way to give pain to boys in their early teens. A couple of stalwarts would hold the lad's shoulders while another would insert a twist of sawgrass, the serrated edge of which will cut bare legs, down the boy's nose. When it was far enough down, the man would pull up the twist and slice off the mucous membrane of the lad's nasal passage. The withdrawn grass would be covered with bloody mucous. The idea behind this and most other puberty ceremonies is that the young fellow needs to suffer to cross over into manhood.

The feminist quickly responded, "And what do they do to girls in that tribe?"

I said honestly, "Nothing that I know of, at least nothing so dramatic." I considered my next reply for a moment but decided what the hell, I might as well go on with her to the end. "Generally, tribal people reserve their most painful puberty treatment for boys."

Whereupon, "I don't think that's fair."

"I never said it was fair. I just said that's what happened." Then the obvious contradiction in her position became apparent. "You mean you'd like those people to torture the girls like they do the boys? It seems to me this is one time the girls lucked out."

Suddenly it became obvious to her. She thought for a moment but then responded true to pattern the best she could. "Well not that exactly. But I think girls should have the same rights as boys."

I did not answer. The lady was a little easier on me the rest of the term

Even more memorable than the troublemakers were the complete oddballs. I remember one from way back, from my teaching days at George Washington University. At the time I was teaching only one class, at night, since I was working as a researcher full-time during the day. It was the standard introductory cultural anthropology course and I was then lecturing on magic and religion. After finishing the lecture on that particular night I was gathering up my books, notes, etc. at about 10 p.m. The classroom was almost empty and I was aware of the stillness in the building. I also noticed a student waiting just inside the door. He was a swarthy fellow who seemed to have some Mediterranean ancestry.

When I reached him, he waited to be addressed. I, "Hello, can I do anything for you?"

He spoke in a low voice, shifting his eyes periodically. "Could I talk to you about something?"

"Sure, go ahead."

He hesitated, then smiled slightly. "It's something different."

I was tired and anxious to be going. Perhaps I was a little impatient. "Go ahead, it's alright."

Hesitation, then, "You'll probably laugh at me."

I went on the alert. This was not a student about to tell me what an excellent lecturer I was or ask me why he got a "C" on his midterm. I, "No, go ahead, I can assure you I have heard enough unusual things that I will not laugh at you."

A long hesitation, then, "You were lecturing tonight about magic and witchcraft. I'm very interested in that."

"Fine, fine. It's standard in the anthropological treatment of the supernatural."

When he spoke again, I had the feeling that his eyes were smoldering. He, "You probably won't believe what I'm going to tell you."

I was very aware of the stillness in the building then and I must confess that I was getting a little nervous. "If you want to tell me, go ahead. If you don't, that's alright too."

Hesitation, eyes smoldering, then, "Tonight you lectured

about witchcraft. Well, I'm a witch.'' I waited to let it soak in, then, "What do you mean? Do you mean you practice witchcraft?"

He did not answer directly but then, "I have."

"You mean you bewitched someone?"

Pause, then eyes smoldering, "I killed a man."

That was news. I began to wonder if I didn't have a looney on my hands. Ah, the trials of a dedicated professor, I thought. But true to my anthropological training, I played it cool, continuing, "What do you mean? Did you use an instrument, a tool?" I was reluctant to say "weapon."

Eyes fixed, still smoldering, "No, I used brain power."

"How's that?"

"I wished him dead, very strongly."

I breathed an inner sigh of relief. For a moment I had thought I might have a materialistic murderer on my hands. This I could understand, perhaps even handle. After all, there are a lot of people around who wish someone dead. So despite his continuing seriousness, I felt the tension lessening. I asked, "Was there a medical diagnosis of his death?"

He looked at me unwaveringly, "They said it was heart failure." He added quickly, "But I know it was my mind power. I have so much it frightens me."

"Okay." I started to gather up my things again.

He, "I was dead once." Adding, "I went to the afterworld"

This fellow was certainly not about to give up. I went along once more, however. "What was it like, the afterworld, I mean?"

"It was like a long dark tunnel with a light at the end."

I didn't give him much credit for imagination on that one. In fact, I think I bordered on being flippant in my response. "It's obvious that you came back."

I think I thought that after all I wasn't then a field worker, collecting ghost stories from native informants. Also I hadn't eaten yet, it was late, and I had almost an hour's drive to get home.

He, "I think you don't believe me."

I got a response back quickly, "Sure, I believe you. I have no reason to think you don't believe these things."

"But you don't think they're true, that they really happened."

Apart from our low voices, the building was completely quiet by this time. Although the conversation seemed to have cooled off, I still felt a little unease. Anyway I said, "It doesn't matter if I think it's true or not. If you say it, it's true that you believe it."

We were both playing word games. He could be pulling my leg the whole time and I was side-stepping the issue of the objective reality of his account. Anyway, he wasn't ready to drop the issue. "I could prove to you that I have all this power."

It seemed to me then that his eyes became more brilliant. I wavered between scoffing and picking up the ball. I opted for the latter. "How could you do that?"

"I could give you a demonstration of my mind power." Pause, "Right now."

The tensiometer needle went up again. I had a fleeting thought of being struck dead in my tracks. But I had the ball and decided to continue playing. "What could you do?"

"Anything you want. You tell me."

I suddenly had the proverbial sinking feeling but decided to go on, to bring the affair to a close. I always had a pressing need for closure. Anyway, I decided to ask for something not too potentially dangerous. "Can you do levitation through mind control?"

He looked puzzled.

"I mean, can you make things defy the law of gravity, float on air?"

He thought for a moment, then, "Oh sure. I can do anything."

"Well, why don't you make those chairs in the back of the room levitate, float in the air?"

He looked at the chairs, then, "I'll have to concentrate."

"That's okay. Go ahead."

He kept his eyes fastened on the chairs for a bit, then turned back, saying, "This needs preparation. I'll have to work up to it. Is it okay if I do it in a day or so? I'll let you know when in advance."

I relaxed then. Whatever else he was, his powers were limited, if existent. "Sure, that's fine. Just let me know ahead of time." I couldn't keep from following one more line of inquiry because of my anthropological background. After all, I had spent much time in the field, collecting stories of ghosts and magic. Furthermore, the student in front of me didn't look like an Anglo-American. I said, "Did you ever hear stories about magic or witchcraft when you were young?"

He thought about that for a moment, then, "Well, my grandmother, who was Syrian, used to talk about people being bewitched." He quickly added, "But that doesn't have anything to do with my power. I really have that."

Where he was really at I will never know. I went to the registration office the next day to see what I could find out. His records told me nothing unusual. However, he never came back to class.

There are many special problems in dealing with a live audience, as public performers of all kinds know well. We professors learn little of such in our formal training. Most of what we get is from direct experience. There have been few efforts to correct this deficiency for college professors, unlike training for primary and secondary school teachers. But of course what medical schools teach doctors how to deal with patients and what law schools teach lawyers how to manipulate juries? Anyway, those of us who succeed in teaching college classes learn on our feet.

One of the ways I have learned for keeping students awake was to use personal experiences. And of course this served another function, ego gratification. And though the ideal may be of the teacher whose sole concern is to pass on knowledge, the real fact is that the most interesting person in the world is oneself. So to be paid to talk about this important personage is a boon which few professions offer. Those who fail to take advantage of such an opportunity must be few. But also examples from personal experience can be just as instructive as any others, especially in the behavioral sciences. For instance, I have taught a course called "The Cultural Animal" which is basically what we can learn of human nature by studying the non-human primates. So when I want to illustrate how primates are social animals and that the males in some groups take a protective role for members of their troop, I tell my Nairobi baboon story. In that instance I got slugged by a big male baboon for getting too close to one of the group's young.

Also in anthropology one has the opportunity of using cultural exotica, which is sometimes relevant, and almost always interesting. When I lecture about eating habits, a most crucial part of any culture, I can bring up some of the food habits and taboos of

Anglo-American culture. The normal red-blooded American is heavily addicted to eating meat and uses most vegetables as side dishes or garnishes, washed down with copious amounts of cow's milk. However, he will not eat most land invertebrates (insects), quite a few species of mammals, including other humans (anthropophagy), clay (geophagy), or feces (coprophagy). In the lectures on language I get to use some of the four letter words (Anglo-Saxon derived) to contrast them to the multi-syllabic circumlocutions (Latin derived). In fact I usually make the point in introductory classes that if students get nothing else, they can at least get a fair line of patter of anthropological oddities for party talk.

I have described at considerable length the good things I have got from professoring, and hopefully that some good things were passed on. However, one important item remains. I am convinced that being a professor has helped me a great deal in maintaining youthfulness, partly in body, but more so in mind. As the years have passed, I have seen the world change over and over. I now have great difficulty imagining the days of my youth, the thirties, not so much that it was quite a while ago; but more, I think, that as the world has turned, I have turned with it. And most importantly I have been able to see much of it through the eyes of my students and other younger people I have known. That has prevented me from locking onto a particular period, particularly that of my youth or early adulthood. Neither of those were a golden age in remembrance. Little nostalgia here.

I have felt more comfortable with younger people as the years have passed. In a sense I see myself as being something of a vampire, sucking not blood, but youthfulness.

Along with anthropologists generally, I know that aging in traditional cultures is a process in which increasing deference is bestowed on the aging. But I also know that conditions have changed in this regard, and particularly in the West. Age is rarely given deference now. Youth is the time to be. And thus I think I am particularly fortunate, since though my youth was mostly a barren wasteland, my later years have more than made up for it. And professoring has been a prime component.

4

Dealing at the Convention

In a social group there are fairly clear-cut rules to accomplish many, if not most, goals. In order to get a longtime partner of the opposite sex, we are expected to get married, and in order to accumulate wealth we are expected to work hard and save money. But many of us, and frequently a large proportion, follow less traditional procedures. We get a partner of the opposite sex by living together and we get riches by manipulating the money of others in illegal or quasi-legal ways, and not necessarily by working hard.

There are informal ways to achieve career goals also, one of the highly important tasks for most, and particularly those in the professions. I have thought long about my career achievements as an anthropologist. And it seems to me that now after a long and satisfying career, that the informal procedures have been as important as the formal ones. And not only that, it now seems that a special occasion was critical, the national convention. Without a number of events having taken place in the American Anthropological Association annual convention, I would never have got to where I did, and might even well not have continued as an anthropologist. Moreover, the most critical events took place in the hallways, not in the formal presentations.

I must point out that I consider myself an average professional, nothing extraordinary, but not of the dregs either. I have worked for about forty years, becoming a full professor about twenty years ago, and have achieved honorable semi-retirement six years ago. But most importantly, I have been completely content with my profession. Some of my jobs could have been better, but there was never any doubt that I wanted to be anything other than an anthropologist. So I am assuming that I am typical. And not just because I am in anthropology. I suspect that my case is similar to that of many others in the other academic fields.

My first significant experience at a convention took place after my most difficult transition as a professional. Either because I was not the best student or because I was too anxious, or a combination of the two, I failed my comprehensive examinations and was thrown out of graduate school. This was at Columbia University, where I was working for my Ph. D. I still remember vividly that traumatic experience when I was ushered in to see the chairman who must have had little to do with my fall from grace but because of his office had to fend me off. Professor Power (a pseudonym) was visibly shaking as he reiterated the news. I remember thinking, "Whatever will I do after all these years? I don't know anything but anthropology. Will I have to begin pumping gas?" In those days most gas was still being pumped by attendants.

Anyway, after it became clear that there was not to be any reconsideration of the exams, I asked in desperation if there was any possibility for a reconsideration in the future. Whereupon Professor Power made what I think was a tactical error. He said with nervous tension (I later learned that he was an alcoholic and I assume that he took a belt or so before facing irate students.), "If you can establish yourself in the professional world on your own, we might reconsider your case."

At the time this hardly seemed any help. How could I go forth and get an academic job or any other in anthropology without a single letter of recommendation from my university? Anyway, nothing I said caused him to change, even though he kept squirming miserably. When I became a professor I learned what an unpleasant job it was to tell students they had failed, especially the ones who were really committed. But when I was sitting in Professor Power's office, all I could think of was how to get through this. No brilliant idea emerged and I finally left, thoroughly depressed.

Fortunately, the same week I had taken the comprehensive

exams, I had also taken a civil service exam for a job in a museum as an anthropologist. This was in Milwaukee, which prided itself on the rectitude of its administration. Civil service jobs were to be handed out precisely according to the scores applicants made on the tests, which had to be given in major universities all over the country. I had taken the test at Columbia without knowing anything about the city administration or the museum. I was not long at the museum before I learned that a local young man had been wanted but his test score was not as high as mine.

Anyway, while I was in the doldrums of my exam failure, notification came from Milwaukee that I was to be hired. I was jubilant, of course, because for this job I did not even need letters of recommendation. I settled into the new job, they knowing nothing of my academic failure and I of course studiously keeping my mouth shut. It turned out that my immediate boss was also from Columbia, though he of course had not been kicked out. And so I was saved, at least in the short run.

I was well enough established at my job when a Fulbright Scholarship notice came across my desk. My area of specialization at Columbia had been India and my boss suggested that I apply for that country. I did, and after several months of negotiations, I was awarded the grant. Needless to say, I had requested no letters of recommendation from Columbia. Fortunately, by this time I could get letters from people at the museum, especially my boss and the museum director, who was also an anthropologist.

I went to India and did a field study, probably not much worse than most beginning efforts. Needless to say, I got no help or encouragement from Columbia, since they did not even know I had gone. I did get a few encouraging letters from my boss and the director, but for the guts of the field study I was on my own.

When I returned, I began working up my field notes with the intention of producing a publication. I had learned that it was generally expected that someone who had done a major field trip would publish on it, especially early in one's career. The museum had been generous to me on the trip, continuing my salary while I was there and giving me time to write my "ethnography" when I came back. This had enabled me to take my wife and son along, since the stipend for the Fulbright student grant had not been enough for three people. And in about a year I had finished the document and had it stashed in my desk drawer.

While in India I had also become fascinated with the marital advertisements in the English language newspapers. The qualities for potential spouses were quite different from those in Western ads, specifying many more social characteristics, as was appropriate for a very socially oriented society. There was practically nothing about love. (See "The Evolution of Presenting Behavior.") Anyway, my boss recommended that I analyze them and write a scientific paper.

So I did my first scholarly paper and submitted it to the American Anthropological Association Convention. That was 1953, at which time the membership was much smaller. Presumably it was easier to get papers accepted. Anyway, they accepted my paper on "Marital Advertisements in India" and put me on the program. I do not remember such as being part of any grand strategy. I was simply doing what I could under the circumstances.

Anyway, by that time I had given a few public talks at the museum and presumably did adequately at the podium of my little session at the convention. As best I remember, the audience was small and no-one from Columbia was in it. However, afterwards I went outside and met the new chairman of the department. I had had one or two classes from him during my student days and remembered him as a nice guy. So after the greeting niceties, I told him what I had done and how this seemed to fulfill Professor Power's requirement for a reconsideration. Professor Wright (a pseudonym) agreed and promised me that he would take up the matter with the committee when he returned to Columbia. In a few weeks I got a telegram to this effect. I would have to take the exams again, but if I passed them, I could go on to do my dissertation. After cramming for six months, I took the exams again and passed.

The problem then was what to do about the dissertation which was already written. Of course there had not been any professorial supervision. I certainly couldn't send it off immediately, so I waited for what I thought was a reasonable length of time, then sent it. They must have thought I was a real fast dude. But in any event after a couple of minor revisions which were suggested by my newly appointed committee chairman, Professor Plattsburgh (a pseudonym), and a tense defense committee meeting, I was approved, and in due time got my Ph.D. No-one at the Milwaukee Museum ever knew anything about this event. But if it had not occurred, by being precipitated at the A.A.A. Convention, I would probably have been found out and my career as an anthropologist drastically changed, if not ended.

Nothing of any great significance convention-wise occurred for several years while I kept cataloging specimens and placing them in cases to be viewed by the public. Then I got the itch to do something for my fellow man, and as an anthropologist, this meant nonwestern man. I suspect that I also had got the itch to travel to some faraway places, a particular disease of anthropologists. So at the next A.A.A. convention I looked around. Now I truthfully do not remember what I did exactly. Perhaps I checked into the listings at the "slave market," the employment agency that operated at each of the meetings. If not, I must have simply passed the word around that I was in the market for change, and particularly for a position in applied anthropology, a new field that had expanded in the last few years.

It so happened that a bureaucratic wheeler-dealer who was well ensconced in AID, a subdivision of the U.S. State Department, was convinced that anthropologists would be useful in village development. The standard types in such projects were technical specialists and administrators. Now it seemed that those in the know were also giving some consideration to cultural factors, the customs of the target people. Anyway I decided to try it, and before you could say "techno-ethnocentrus," I had signed up for a two year stint, doing village development work in Laos. It would be devious of me if I did not also mention that I would be getting a considerable raise in salary, plus quite a few other perks for myself and family (self, wife and son). But I honestly thought I would be helping some few of that primary underprivileged group of the world, the villagers.

Unfortunately, I had learned little of the "realpolitik" which controls international relations. AID, being an arm of the State Department, was designed to foster the interests of the U.S., and if helping villagers to grow more food, or get better education or health, would further U.S. ends, no objection. But other means might be deemed more appropriate, major ones of which were politics and war. In those days the U.S. was very much against Communism and in favor of governments controlled by the people, whatever that means.

And when people of other nations did not willingly accept these ideas, along with some other factors, the U.S. government was not unwilling to call in the military, their own and that of the "friendlies," (read non-Communists). And so in Laos the village development scheme played a distinct second fiddle to diplomatic and military intervention. I rapidly became frustrated, so that by the time my two-year stint was over, I was unwilling to do more. And though belated, my assessment was right. The war escalated very rapidly after I left and no-one will ever know how many peasant villagers were blasted out of existence. And though I know this has happened to the peasants of the world for the last several thousand years, I still feel badly about my involvement in that poor country.

But life goes on and mine did also. I came back to the U.S. without a job and with the same three mouths to feed. Where to go to get on track again in anthropology? I stopped by the old university where I was by that time in good graces. The word by everyone was that I should get a university teaching position. I was not getting any younger, I was told, and if I did not make the shift in the next 5-10 years, it would probably be too late. I was 41 years old then.

The most useful thing I got out of my Columbia visit was a get-together and subsequent agreement with Professor Plattsburgh to write a handbook on community development for American technicians. I did it that summer and fall.

Then it was back to the "slave market" at the A.A.A. convention again. This was 1962, universities were expanding, and there were many academic jobs around. Within four days I had several nibbles and two offers. I took one at the University of Wisconsin-Milwaukee and was shortly in the classroom -- in the middle of the northern winter. And though my appointment there was initially temporary, once I got started, they offered to keep me on. However, I just could not resign myself to stay through those winters again. So after a year and a half, I headed again for the A.A.A. convention. This time I did not have to hunker down to the "slave market," since my book, co-authored with Professor Plattsburgh, had been printed in an academic edition and I had a minor reputation as an applied anthropologist. There were even people who looked me up. And among them was a guy who was building up a research group in Washington, D.C. It seems that then, the early sixties, there were besides many academic positions, also many expanding research groups. And to get a cooperative anthropologist on board these was the thing to do. The guy who hired me, as well as most of my colleagues for the next eight years, was from psychology, the popular human science. And so I studied village development projects, eventually to publish the results. Once I was settled in Washington, I began teaching one course on the side, most often at American University.

My chief at HumRRO (Human Resources Research Office) continued to look for more anthropologists after I was hired. And so I met Black Bart. He was an anthropologist at a small university and on the lookout. As best I remember, my chief found Bart at the A.A.A. convention. Bart got his name, I think, because he was a gun-slinger, academic that is. He usually fired fast and asked questions later--more on that later. Needless to say, he had another formal name. He spoke very softly but as I learned later, this was no indication that he was a gentle, retiring person. When he talked to you he usually had a slight smile as if he knew something you didn't. Little did I know at that first meeting how much he would affect the rest of my life.

My chief, who we referred to as the Inscrutable Occidental because you could never tell what he was thinking about from his expression, indicated that he was less than impressed by Black Bart, even if he was an anthropologist. In other words, no job offer was forthcoming. It didn't matter much to me, nor to Black Bart for that matter. Presumably he had tossed out the bait many times at the A.A.A. convention in time-honored fashion. I simply continued with my study of village projects.

But for whatever reason, Bart did not forget me. After many months, perhaps even a year or so, I got a call from him. He was then chairman of the anthropology department at one of the California State Universities. He must have got the position when he had been fishing, a part of which had been his interview at my research unit. Moreover, he offered me a faculty position. I was ready, had in fact continued to prospect on my own for several years. Although I liked Washington, D.C., there were a number of negatives. Also I still remembered the admonition of my professors at Columbia to get established in academia before I got too old. So I agreed to come to

California for a look-see for myself and for them. This was favorable and in a few months I was at California as a professor, specializing in applied anthropology.

I settled down to teaching and quickly realized that I enjoyed it. However, it took me even less time to realize that Black Bart was having troubles as chairman. He had alienated a lot of people, including well established deans, old faculty members and politically adept students. It seems the soft-voiced gunslinger was a very opinionated guy who expected to impose his opinions on anyone he came in contact with. In less than a year after I arrived, the department had been given an ultimatum --dump Black Bart or be merged into the Sociology Department. The solution that was worked out was for the original founder of the department, a guy almost anyone could get along with, to step in as chairman and to get a regular after that year. The regular was of course me. Within one year after I came, I was department chairman.

And though this essay is about dealing at the A.A.A. convention, not personalities, I simply cannot drop Black Bart without some explanation as to what happened to him, if for no other reason, the accidental conversation I had at an A.A.A. convention, some twenty years later. I was having a drink at a table in one of the expansive lobbies of the convention hotel. Another guy with a name tag sat down next to me. His tag read Seneca University. I did not recognize his name. I said, "How are you finding the convention?"

"Oh, standard."

I kept on, "I had a connection with your university once."

"How's that?" he asked.

"One of our faculty members took a position at your university, guy by the name of Bertram Goldberg."

He snapped to attention and his facial expression changed as if he had been hit with a wet towel. He peered then at my name tag and said almost with a touch of anger, "Yes, I see, you are from California State University. You sent us that guy."

It was about twenty years after all this occurred and I was amazed at how much emotion the name of Black Bart could still evoke. He certainly had a special talent for stirring people up. I knew more or less what had happened from accounts of my colleagues of that time. When he had been unceremoniously demoted from being department chairman, he had got on the phone to Seneca University, on whose faculty he had a friend. And through how much and what kind of cajoling, promises and other persuasion one will never know, he had convinced them to hire him as department chairman. And from what I heard, he got about as many perks as was possible, including full professorship, tenure and a good salary raise. Not only that, I don't think any references of his record at our university were asked for or sent. Within the next year reports began to come back that Black Bart was repeating his performance there and shortly he was deposed there also, permitted to keep his teaching position but forbidden to take part in any administrative activities. He stayed on at the university for a few years, meeting his classes, but remaining scarce otherwise. Finally he resigned and went off into the world of small business, ending up in real estate where he was reported to be successful.

Through the years I met different people from Seneca University and they invariably accused me and my department of sending them Black Bart. I invariably replied that no-one there had asked any of us for a reference when they hired him. And so I did the same with the last guy, maintaining that any problems they had with Black Bart was their fault.

Anyway, after he was gone, I remained at California State University as chairman, a position I stayed in for eight years with a couple of years off on leave. This was the early 70's and universities, including ours, were still expanding. So for the next 5-6 years one of my primary jobs was to hire new faculty. Where to go but the "slave market" where I and so many others had got started. Year after year I registered in the employer section and for four hectic days tried to interview as many applicants as possible. Everyone in the department now who is older than 45 I hired during that period. I remember being continually harassed those years at the convention. I would survey the applicant files first, make up notes for likely candidates, then pick up notes left for me by hopeful applicants who had read my listing. But also I was continually being stopped in the hallways by prospective employees or others who were passing the word around. There were also phone calls in my room or notes in my hotel key box. Then if all bode well, I would interview the candidates. Some of the people I hired still joke about my interviews. One pair who are still teaching in the department laugh about their interviews which were together. They claimed that after a brief conversation with me, one of them asked when the interview would take place and was told that it had already happened. But no matter how hurried or harried I was, most of the hirings that I did, including these two, turned out well. Both men have served as chairman of the department and produced respectable publication lists, as well as teaching many successful classes.

I hunkered down then as chairman and professor and for several years did little else professionally. But then finally I got finished and turned to my first love, writing. I had always wished to do creative writing and though I had buckled down to do stuff in scientific/academic style when I was on my climb up the academic ladder, once established as a long-time professor and ex-chairman, I turned to popular writing, although with a heavy tint of anthropology. I could hardly think without referring to some concepts from the field.

I wrote essay after essay in popular style until I got to one which wouldn't stop growing. It derived from a main interest of mine, cultural change. And true to my anthropological training, I took the natives' point of view. It grew into a book, I rewrote it a couple of times, and it was finally published. It was written in a science fiction format. Though the story line was of an old anthropologist being interviewed by aliens who had conquered the earth, the subject matter was a description of the domination by European man over the other peoples of the earth. The title was "Another Side of History: What Euroman Did to the Natives."

I was very elated that the book had been written and published in a popular style. I even had fantasies that it might be popular. Such was not the case. Most people who read it, claimed that it was quite good. Moreover, it seemed to me that it was timely since it came out just the year before the Columbus Centennial. However, the book was never reviewed, picked up by any booksellers, or sold well through mail orders. I got temporarily depressed.

What to do? As usual, my thoughts turned to the A.A.A. convention. My publisher had never done a book in anthropology before so he had done nothing in advertising the book. Therefore I took several copies to the next convention, and finding that there were no sample copies in the book exhibit, arranged at two multiple publisher exhibits to have copies of my book on display. I was temporarily elated to hear at the end of the conference that my book was one of the most popular ones there. Evidently some anthropologists liked it.

With my temporary "high," I took other measures to publicize the book, but with indifferent success. Obviously, it was not "grabbing" people as I had hoped. And whether it ever will is an open question. But so far as this essay is concerned, the publication exhibit at the A.A.A. convention gave me a needed stimulus to go on.

There is a term nowadays for the kind of encounters I have described. which is networking. It is a good example of the Rumpelstiltskin effect, which is that something doesn't exist until it is given a name. I was doing it long before it got a name, which was being pragmatic.

5

The Publishing Game

Anyone who has been in academia has played this one. To teach in a university one now needs to have a Ph.D., which includes having finished a dissertation, a written work of 100 + pages, based on observational study (first or second hand). One does not necessarily need to have published, though that certainly doesn't hurt. Some of the prestige universities have required that the Ph.D. candidate have his dissertation printed, at his expense of course. A continuing requirement is that the work be done under the supervision of the professors. (See "Dealing at the Convention"). And though there have to be others, I know well how one widely known anthropologist, Carlos Castenada, had a running battle with his anthropology department for months or years because the work he wanted to use for his dissertation had been published commercially and became a best seller. His university department finally accepted it.

Anyway, our hypothetical professor has now snagged a position in a university. He knows that to become tenured and to advance beyond the rank of an instructor or assistant professor, he is going to have to start producing printed words. It doesn't matter so much how important his works are, but there have to be so many of them. During my regular period as a professor (tenured and full rank), I was on a college promotion committee which came to an impasse about what to do with professors seeking promotions who had only published papers proving a "null hypothesis." This is to prove that a hypothesis, read "hunch," was wrong. It was typical of professors in the hard sciences such as chemistry, physics and biology. It also involved the researcher having duplicated some experiment, usually of somebody else. It is a technique that makes some sense, based on the built-in scepticism of the natural sciences. The normal human inclination is to want one's hunches to be correct. But the hard scientists have built their reputation by requiring multiple proofs by independent researchers of a hypothesis. They always know there is some young faculty member trying to climb his way up the rank ladder, and who needs publications on his vita. So they are more careful about their proofs and try harder not to foist off some untenable hypothesis. Needless to say, anthropology is such a loose bundle of hypotheses, theories and methods that the "null hypothesis" procedure is hardly relevant. But lest the reader think I am grossly disparaging anthropology, let him know also that I believe the "science of culture" has many other positive aspects. Anyway on the aforementioned promotions committee, we decided to give young faculty members full credit for publications based on the "null hypothesis" procedure.

So our hypothetical faculty person got his appointment, and then by producing published papers and books got his tenure and moved up the rank scale to full professor. I might mention that tenure is an academic status, the acquisition of which makes it very difficult for the administration to get rid of the person (See "Dealing at the Convention"). But will he really perish if he doesn't get tenure? Not exactly, at least if he has his Ph.D. The usual procedure is to simply not renew his contract after 3-5 years of no publications. Then he looks around and finds another position, most frequently in a lower status university. We did this with several persons during my chairmanship. Most got jobs in smaller or lower paying schools or in ones which were in remote locations.

Once he has full tenure and rises in rank, our hypothetical faculty member can lean back, and if he does not have much of an incentive otherwise for publishing, he can pretty much drop it. This will depend to a large extent on the status level of his university. The more prestigious it is, the more it will expect its professors to continue publishing. But even in those of highest class it is unlikely that they will be dumped if they discontinue publishing after they have tenure and full professorship. So the widely quoted "publish or perish" should properly be "publish or perish until you get tenure and/or full professorship."

Some faculty types have the good fortune in that they really like to see themselves in print, either for disseminating information, vanity or profit. I suspect that vanity is the most common motivation, disseminating information would be second and profit last. Except for a strict minority each generation, most writers in academe make little money in writing, and those are the popular textbook writers, a handful. I happen to belong to the smallest group, because even now in mandatory retirement I get my rocks off writing about the subject. Even when I was doing standard academic writing, however, I often enjoyed it. And I did go through the usual academic writing procedure, to establish myself and climb to the rank of tenured professor.

But as has occurred to me in other endeavors, my achievements in writing have sometimes been offbeat. But they worked, providing me with a thoroughly satisfactory profession. If given the chance for another life and another profession, I would probably take the former but remain an anthropologist the second time around. Oh yes, there is one other profession I might have taken -- that of a creative writer. But if so, I would at least insist that anthropology still be the primary basis for my thoughts -- and writing.

The main story here is how I, with indispensable help, took on the U.S. State Department and U.S. Printing Office -- and won. I always did want to put that story down in written form.

But first I need to fill in the years before Steve Boggs, executive secretary of the American Anthropological Association, and I launched our campaign. As described earlier, from my late teens on, I wanted to be a creative writer, to write stories. But when I went into anthropology that had to be put aside for a while. But in my first job at the Milwaukee Public Museum I was lucky enough to get a chance to do both academic and creative writing. They had a publication series in anthropology and encouraged me to publish in it. Needless to say, I did not have a reputation at that time that caused publishers to come banging at my door for my productions. So I published two ethnographies in the museum series, including my

Ph.D. dissertation. These were standard anthropological works, probably a thousand copies printed, most of which went to libraries and area specialists. I have seen them quoted in years gone by but they have long been out of print.

More important for my need to do creative writing were pieces I did for the popular museum journal, <u>Lore</u>. It was similar to the popular magazines of other large museums such as <u>Natural History</u> and the <u>Smithsonian</u>, a brainchild of our museum director, W.C.Mckern, to bring museum lore to the public. "Mac" really doted on that magazine and he encouraged me to publish articles in it almost as soon as I was settled in my new job. So for some eight years I did articles for <u>Lore</u> and enjoyed doing every one of them. I became curious recently about some of these and requested one "The Sorcery Book Publishers" from the current curator. I was honestly surprised at how well it read and have thus included an edited version in this work.

During my museum period I also got accepted and published three or four articles in professional journals. The most significant one was "Marriage Advertisements in India" in the <u>Indian Anthropologist</u>, the study I did which enabled me to ultimately get my Ph.D. ("Dealing at the Convention").

For three years after this, I did little writing except reports, because I had joined AID and was working on village development in Laos. Then in disillusion I resigned and came back to the U.S. In less than a year I had become an academic. In that interim, however, I made arrangements with my old professor, Conrad Arensberg, to rewrite a manual for development technicians. The American Anthropological Association had been given a contract by AID, the foreign aid section of the State Department, to do a practical work to help guide technical specialists through the cross-cultural mazes they faced overseas. The manual had been written two or three times already and rejected by AID officials, each time on the grounds that it wasn't practical enough. Then the assignment had been handed over to Arensberg, since he was known as an applied anthropologist. He was very busy, and when I showed up after my stint in Laos he offered it to me. We made an arrangement whereby I would rewrite the rejected manual but then we would publish it under both names. This worked out well and I did it in a few months. The manuscript was accepted and handed over to the government printing office. In the meantime, after a year and a half teaching in the Midwest, I took a job at a research agency in Washington, D.C., HumRRO. My boss there asked me to edit the new manual on company time so he could get a number to pass out to the technical specialists whose agency was paying for our group. That was the U.S. Army, which at that time had a program called Civic Action. It was supposed to be helping the villagers of the world, or at least those in the countries whose policies the State Department approved. As usual with State Department policies and programs, this one hardly accomplished its task. But in the meantime I cleaned up the manual and kept telling myself that this might help some peasants, no matter the auspices. And so the manual,"Technical Change and Cultural Reality" got printed, and shortly 100 copies which I had requested, were sent to me. By this time I had found my way around Washington and had got to know quite a few people in anthropology and development. In particular, I had got to know Steve Boggs, the executive secretary of the American Anthropological Association. He knew of the history of the manual and was pleased with the current development, since he was interested in tieing anthropology in to government.

Then one day I got a call from an AID official who requested that I hold up distribution of the 100 copies on the grounds that the publication had not yet been cleared by whoever did such clearing. He could give no logical explanation why the book had not been approved except that some of the examples put Americans overseas in a bad light. One in particular that he cited was an introduction of a new variety of corn in South America, which because it was much harder than the traditional variety, had been used by the villagers to make beer. I tried to tell him that many of the varieties of the natives were used for that purpose and that corn beer was one of their staples. But he was in no mood for a mini-lecture in anthropology and this got me nowhere. But he said I needed to keep the books for a few days only. It was only a formality, he said. I told him I would think about it.

I called Steve Boggs and told him what happened. We agreed that we would wait for a week or so. I called the guy back the next week and was told that the approval had not yet been made. Everyone over there was so busy! I waited another week or so and got the same answer. Whereupon Steve and I decided that we had kept our agreement each time but that the prospects didn't look so good and that our only hope was to get the hundred copies to anthropologists who might help.. After all, the manual had been done through a formal contract with the American Anthropological Association. We drew up a list then of the chief applied anthropologists we knew in academia and mailed the copies out. I stopped calling the AID spokesman and the best I remember he never called me back again. Then for the next year I answered letters from anthropologists who wrote to ask for copies to use in their classes. I worked out a regular form letter in which I explained what happened and requested that the writer use his influence in government circles if he could.

A request came through from Professor Sol Tax of the University of Chicago who was a well known applied anthropologist. I sent him the standard reply. He answered back that within a few weeks he was coming to Washington on a business trip and he would see what he could do. He got an official from the United Nations to accompany him to an appointment with the director of AID. Neither Steve nor I were at that meeting so we had to use our imaginations and what Tax told us on the phone. It seems the director was anxious to have good relations with the UN. Furthermore my guess is that the manual was a peanut affair in comparison to all the more ponderous decisions that were being made, mainly which countries were going to get how many millions/billions of dollars and which were going to be shut out of the cornucopia. Steve and I could never figure out what was subversive about the manual anyway. None of the examples that had been cited were presented as those of AID. The "corn" one wasn't even my example, having been taken from the previously written version.

The main problem is that Washington bureaucrats get nervous when appropriations time comes. I suspect some middle level official had visions of a congressman getting hold of a copy. Anyway, the top dog, the director, with a little influence from Sol Tax and the unknown UN functionary, cleared the manual on the spot. One of the most bizarre aspects of this whole affair is that it was done entirely through the mail and phone. I never saw a face of one of the principals except Tax, who I saw at later A.A.A. meetings. I do not recall that we ever discussed it face to face. And that is how we beat the U.S. State Department.

The manual was sent to American missions overseas, primarily to AID technicians and Peace Corps Volunteers. And the story would have ended there if there had not been one unexpected fall-out. Just shortly after the AID clearance, I got a phone call from a publisher in Chicago. On Tax's recommendation, he was prepared to publish it in a commercial edition. The manual, having been done in the government printing office, was in the public domain. Anybody could reproduce it for free. But Alex Morin, the editor of the Aldine Publishing Company, knew that if one changed such a work in any way, one could copyright that version. So we changed the title to "Introducing Social Change," put in an introduction, and made some minor editorial changes. It went into print and still is, 34 years later.

Anyway the book got around and what reputation I have as an anthropologist is primarily because of it. I am sure I got solidly into academia on the basis of it and there were some other consequences. There was also a spin-off in another book, "A Casebook of Social Change" which did okay for 10-15 years. Also I got some articles in applied anthropology published as a consequence.

My major activity for the next eight years was in teaching and administration (chairmanship). When that was over and a new chairman appointed, my thoughts went back to my first love, creative writing. I certainly did not have to worry about tenure and rank anymore and could more or less drop academic writing. But before I could get deep in the writing, I had a serious illness, a stroke. When I started to recover in a few weeks, I turned to writing as therapy. I barely got the book done about the stroke when I got hit by one of the most traumatic social illnesses of our times, a divorce. Two more novels for therapy, this time against socio-psychological illness. Then life improved and in no time I was deep into anthropological writing again --starting a long series of essays, practically all with a strong anthropological component. They were in popular style however.

One essay wouldn't quit. It was the acculturation of the world by European man during the last 500 years. Anyway, it got to be book size and went through two rewrites and finally publication. By this time, however, I had amassed some 20 or more essays, all with a strong anthropological component. I decided to entitle the package "Anthropology as a Way of Life" since I had come to the realization that I could think of no subject without calling on some concept or data from anthropology. Then one of the last ones, "Dealing at the Convention" was accepted for publication in the "Anthropology Newsletter." Since then I have attended a number of conventions, but as a retiree, there has been little cause to do any further wheeling and dealing.

6

The Sorcery Book Publishers

    The unpublished author is compelled by his need to get books published to try every method that will offer some hope. The most common method is to send manuscripts to publishers. This doesn't work very well, because publishers are inundated with packages of paper with writing on them. Another method is to try to get an agent. This is also difficult in that agents are not generally interested in writers until after they have already published, a distinct minority of the total number of writers. Some charge a fee merely for reading the manuscripts. Nothing is guaranteed and the unscrupulous go no farther. Then also the author can pay the publisher out of his own pocket to have his book published. These are sometimes called vanity presses. And finally the author can assume the role of publisher himself, thus handling all aspects of his book, including its promotion. This is called independent publishing. It has been resorted to many times, often by men who later came to be widely recognized as accomplished writers.
    As one who has been scribbling fiction during his spare time ever since college days, I have tried all these methods with some, though hardly fantastic, success. Furthermore, I have had one means of getting a book published which very few other authors have had. This was through magic. My opportunity occurred because I am an

anthropologist and in the course of field work have had access to sorcerers.

During a field trip to Trinidad in the West Indies, this opportunity arrived partly by chance and partly by design. The East Indians, whose way of life I was studying, were very much involved in the practice of sorcery. From the time I arrived, I kept hearing stories about sorcerers and seeing evidence of their handiwork. People who were sick or who were believed to be under the influence of evil spirits were taken to sorcerers for curing. People who wanted better jobs, or young people who were trying to get a lover or keep a spouse on the straight and narrow, went to sorcerers for assistance. Men were always on guard for evidence of sorcery against them by their wives. Various charms, provided by sorcerers, would be found hidden under the unfaithful husband's pillow or buried in the yard. If counteraction was not taken, it was believed that the husband would be bound to his wife regardless of his wishes. During my period of research, none of the common reasons for going to a sorcerer occurred to me. I was not looking for a better job, my wife was not unfaithful to me, nor I to her, and I was not sick. However, as part of my research I felt that I should observe, first hand, the technique of at least one sorcerer. With this in mind I made arrangements with some of my East Indian friends to visit a well known practitioner. It should be mentioned that sorcery was against the law and thus the practitioners their activities in secrecy. Thus, it was not easy for an outsider to get to know them well. However, through the intercession of my friends, I was accepted as a client.

This man practiced in a hut about a half mile from the nearest road. In order to reach the place, one had to walk through rice fields and cacao (chocolate) groves, along a path that could only be followed after receiving explicit instructions. A number of people were already waiting when we arrived, and the sorcerer was talking to a client in his thatched hut. Rice bags were hung up as curtains in the entrance doorway. Among them were several modishly dressed women, wearing high heels. To have come through the rice paddies with such shoes would have been a difficult task.

When I and the two Indians who were accompanying me were ushered in the house by the assistant, the sorcerer was seated against the wall with half closed eyes. He appeared to be in a kind of stupor.

Near him were a bottle of rum, a shot glass, some matches, and a package of cigarettes. At his request, I had also brought along a bottle of rum and some cigarettes. As soon as I and my two friends were seated, I was asked to say my name out loud. When I did this, the sorcerer started to repeat it with tight lips. The assistant thereupon poured a jiggerful of rum over the sorcerer's head and gave him two cigarettes which he put in his mouth and lit simultaneously. From this time until the trance was over, the sorcerer smoked two cigarettes at a time. He would light two new ones each time the old ones had burned down. Also two times more a jigger of rum was poured over his head. Without opening his eyes, the sorcerer began to mumble in a barely audible voice. He was obviously speaking English, but whatever meaning there was in his monologue was unintelligible to me. However, I could distinguish the words "clarification " and "fortification"several times. My feeling was that the man was not making any logical statements, but was rather using strings of unrelated words, most of which were blurred beyond recognition. At intervals he would shake his head from one side to another, making a nasal sound like "mmm mmm."

    The assistant interpreted for the sorcerer. I was told that the sorcerer was under the possession of a spirit; also that people were speaking evil about me behind my back; and that I would not achieve success until this "stumbling block" was eliminated. I asked if these people were in Trinidad or in the United States and was informed that they were abroad. I didn't think this was very remarkable information, because I had always believed that it was one of the natural conditions of life to have people speak behind one's back. I was then asked if I had any requests to make to the spirit. I racked my brain for some special request which the assistant said was necessary. Finally, the problem of how to get my books published came to mind. I said that I had a book which I would like to have published. The question didn't seem to register and the assistant told me the sorcerer needed a clarification. I rephrased my statement, saying I had written a book and that now I needed to have it printed. Still there was no direct answer to my question. Finally, I was told I needed clarification on both my problems and that after this was done, success would be assured. In order to do this, I would have to return on Wednesday of the next week. After conversing with the spirit about a problem my

East Indian friend offered, the sorcerer stopped smoking and soon returned to a normal state. We conversed for a short while and made arrangements for our return --for the clarification. Before we left, the sorcerer hospitably offered us a drink of the rum that his assistant had been pouring on his head during the trance.

    I came back on the specified day and found the sorcerer in his house instead of in the hut where he had conducted the trance. He had already prepared part of my "medicine" and set about getting the ingredients that were missing. I was given a leaf of a plant called "bakando," 21 small elliptical leaves of another plant, a charm container, and a bottle with some red liquid in the bottom. The instructions for using this material were as follows: Go to the seashore and fill the bottle with sea water without emptying the red liquid. Take off clothing. Take three leaves from the 21 and holding them between the fingers of the right hand, cup hands together and dip them into the sea, sloshing water over the face. Do this seven times. On the seventh time spit on the leaves three times and let them fall into the water. Repeat this procedure until all the leaves are gone. Get dressed but do not wipe the water off. As soon as clothes are on, turn away from the sea and take "bakando" leaves, to which salt and a pinch of garlic have been added, and pass three times across the bottom of each foot. Hold "bakando" leaf up and say to it, "You have followed me so long and you will follow me no more." Spit on it three times and throw it backwards through the legs without looking at it. The charm had to be worn on a string at the right side of the waist. It was a small metal cylinder, containing two long strips of school tablet paper which were rolled in a cylindrical shape in order to fit into the container. The inscriptions found on them were as follows:

    N I R

    All this be to grant eternity, here in time. Sanctus Spiritus.

    N I R Amen

    Ananiah, Azariah and Mizael, blessed be the Lord, for he has redeemed us from hell, and hath saved us from death and he has redeemed us out of the fiery furnace and has preserved us even in the midst of the fire; in the same manner may it please him the Lord that there be no fire.

    I N I R

    The charm had to be brought back to the sorcerer about a

month later for the invocation. This consisted of passing it around a candle seven times while saying certain prayers. I was surprised to note that the magic formula in the charm was written in English and I asked the sorcerer about this. He said he had Negro, Indian and White spirits and the spirit which he used for me was a White spirit. Also I asked once more about my definite request, that is, when I would find a publisher for my book. Again I was informed that as soon as I was clarified, this would take place.

Some true believers may ask "Did the book get published?' I must report that it did not, though in honesty I did not follow through with the sorcerer's instructions. I did not take the ritual bath, say the incantations, nor wear the charm. As a Western-trained social scientist I tend to rely on naturalistic explanations. Furthermore, I thought I would look ridiculous standing at the edge of the sea undressed, spitting on three leaves, and then throwing them backwards through my legs into the sea.

7

In Praise of Libraries

If I were to dedicate a statue of my own choosing, there would be no doubt as to whom, to Johann Gutenberg, the inventor of movable type and the printed book. Of the many innovations for the storage and transmittal of thought, after writing itself, none compare to the invention of printing. It was the sublime achievement of technology. We are now in the electronic age, and there are those who believe that the digital computer is one of the greatest inventions ever to come along. Though I readily admit its marvelous capabilities, it gives me comfort that the primary message on the electronic screen is what we now call "copy." This is nothing less that the characters which Gutenberg froze in movable type. Also the usual way to learn how to use a computer is with a book.

So in the name of technology, humankind has produced marvels. But as we have so often learned to our chagrin, how we use our technology is just as critical as that we have it. And the way we use it is a product of our social systems. We organize ourselves to build roads, fight wars, worship gods, and live together. So it is inevitable that we would organize ourselves to use our newly found communication artifacts, the printed page and book. Thus, we invented the library, a place where we could store our treasures and from which we could disseminate them.

As anthropologists, we teach that the prime achievement of humankind is culture, a system of ideas which is transmitted with changes from one generation to the next. Furthermore, it is transmitted primarily by means of language. We "goo-goo" to the baby almost immediately. But the "goo-goos" rapidly change, along with the meanings, into the words of our language. Then these are transformed into sentences, and finally into combinations thereof, until we have discourses. In the process we are teaching all the accumulated knowledge that is worth keeping since the generalized cries of the ancestor changed into particular sounds with particular meanings. Thus we got language.

As a species we got by with the spoken medium for a few million years. But then some bright persons decided that we could freeze this speech. They invented writing. We have no idea who that was, it occurred so long ago. And so far as we can reconstruct it, the original motivation was greed -- how to tax and use the lower classes for the benefit of the uppers. This of course is a condition still with us, the tax collector of ancient Sumeria being replaced by the I.R.S. agent, the supreme monarch by the politico.

But writing enabled humankind to record other kinds of information also. It permitted the busy biped, *Homo sapiens*, to write down his folk tales. Thus we got the Gilgamesh, the Odyssey, Beowulf, the Vedas and the Old Testament. In the long run it made organized religion possible. And other kinds of history were also written down, both imaginative and real. Plus knowledge of the universe. Aristotle described what he saw in his limited world, as did the later Arab geographers, and finally Euroman when he "discovered" the planet.

Among other things, this freezing of knowledge made the Renaissance possible and then the development of the scientific way. Of course by then we were in the age of printing. Probably building on the knowledge of the much earlier Chinese experimenters, Gutenberg came up with his marvelous invention in 1450. After that, all knowledge, once written, could be disseminated enormously. We were no longer in the restricted world of the Middle Ages when all knowledge had to be laboriously copied by monks.

But still the ability to read was limited by the educational facilities of the respective societies. Reading and the use of books was still a matter of class membership. The upper classes learned how, or had scribes do it for them. They also had access to a newly invented social institution, the library. This was a place where the printed material of the culture was stored and used by those with the proper class credentials. It certainly did not include the majority, the peasants and working class, because they did not know how to read. They were automatically disenfranchised. The upper classes didn't want the peasants to read, because access to information made them less exploitable. And whatever else they were to the upper classes, the peasants and city workers were exploited people.

The library of course existed in the age of hand copying and even back to when they were still punching characters in clay. I do not know how far back the collecting of books goes, but probably to the beginning of writing itself, back to ancient Sumeria. There must have been collections of cuneiform tablets. How else would one keep tab on devious peasants when tax time came around? Historians have often cited the center of Greek learning as the library of Alexandria. And of course during the Middle Ages, when European knowledge was in the hands of the monks, there had to be many collections of manuscripts. And the Arabs, when they expanded with Islam, and were promoting learning, also set up repositories for collections of manuscripts. Probably the last bastion of restricted learning was the university library.

After that we must jump to the 19th century, when the democratization of learning took place. As royalty and upper class control weakened, there was an increase in the spread of learning, and particularly literacy. "Reading, writing and 'rithmetic" were the pillars of education in America, where upper classes did not get established. And the new idea that the state should establish schools for the public also became a fact of life.

Both my parents, who were born in the last decade of the 19th century, had gone to school in rural Indiana where they learned to read, write and "figger." They had seventh and eighth grade educations and considered compulsory education a norm. During their lifetimes my mother read some books and my father read the newspaper. Their reading choices were limited because they were devout Catholics. They abided by the lists of the Legion of Decency which were posted in the back of the church. I know well because this was one of the main sources of conflict I had with them as my taste for reading grew steadily broader.

Along with the increase in reading came the development of public libraries. After all that time, 5,000 years of writing, most of the knowledge that had existed became available to anyone who could look it up. And even if they couldn't, there was a new kind of person who would help, a librarian. When it comes to transmitting the accumulated knowledge of a culture, who could ask for anything more? This was the cultural way "par excellence."

In my later years I have become disillusioned about most publicly funded programs for the simple reason that they do not work. They do not deliver the promised results and as often as not their unexpected side effects are harmful, if not downright disastrous. So it comes as a real pleasure to regard a social institution, financed by taxpayer dollars, which really does provide what it promises, the public library system. And why is this so in contrast with so many other public institutions? First off, a library system is not one which attracts devious people. What con artist, political or non-political, is going to spend his efforts trying to cheat the users of an institution whose function is to store and loan out reading and other communication material in small units? The library is not the obvious focus of either money or power, even though the ideas therein may in the long run have more effect on individuals and groups than all the financial and political institutions combined. But anyway, devious types do not generally use the library except for getting information. And many are not literate enough for that.

The fact that con artists, political and non-political, do not base their schemes on library collections is why libraries do not make good newspaper copy. Newspapers go for skullduggery and violence, and those can be found where there is much power and/or money, not in a poor library.

Another positive of the library is its people. They are trained to be professionals, as well as being selected and judged according to their ability as librarians. Unlike the politico, who is judged by the voters in getting his job, the librarian is judged by her/his peers for her ability as a librarian, not as a vote-getter. This makes for a high quality specialist, as is a doctor or engineer, or other types who get their position by ability. These specialists cannot afford to let many people die in surgery or buildings fall accidentally. The equivalent in a politico would be to judge him/her as an administrator before allowing any campaigning. The Chinese went in that direction with their civil service system, though unfortunately they got hung up on the teachings of Confucius as the criteria for selection. Anyway, the librarian is a highly qualified specialist, not a popular vote-getter.

Another very special and pleasant characteristic of the library is that it works to give service. One of the galling features of giant institutions, government or commercial, is that even though they are supposed to serve individuals, the individual can often profitably be ignored in favor of the mass. Civil service institutions tend to get that way. A couple which quickly come to mind are the postal service and the department of motor vehicles. In both, patterns have evolved to minimize the individual and to make him/her wait. Also banks tend to be this way. After using the big city versions of these impersonal, money-making institutions most of my life, I have in my later years been very fortunate to live next to a small town where I found a small family bank (Fallbrook National) and a nearby small-small post office (Bonsall). Both are highly personalized. The bank adjusts to client numbers by adding and subtracting tellers rather than making customers wait in lines. The post office clerks generally do the same. In both institutions the clerks know most of the clients and are invariably friendly. But even in the large libraries that I have gone to during my lifetime, polite service was customary, though admittedly the small, local ones have been the best. But in no library do I remember long lines or impolite people.

Since this piece was inspired by a personal recognition of the importance of libraries in my lifetime, I will give a short history. The earliest one I used was a neighborhood public library on the south side of Indianapolis, Indiana. I remember it as a smallish brick structure but to me then containing an astronomical number of books. Of course if I went back now, and if it was still there, it would look minute. I used to go there on a bicycle, strapping the books down on a carrier for the trip home.

I was then a student at a Catholic school. Anyway, for reasons I am not sure of now, I was having considerable trouble at school and consequently at home, since my parents supported the nuns fully. I know that I had learned about the considerable variety of books at the library, some of which I read. The Church posted a list of forbidden books. I don't remember if I read any of them, but I know it became a source of conflict with my parents. Also, I was evidently getting contrary ideas from my reading, which I brought up in class, and for which I was punished by the nuns. The two kinds of punishment I remember were being cracked on the knuckles with a yardstick and being required to stay in the nunnery after school.

I made contact with one of my earliest friends a few years ago, a man who stayed in Indianapolis all his life. His recollection of me at that time, 60-65 years ago, was that he didn't know anyone who read so much. The only explanation I can think of that caused such was that I was pretty much a social misfit, and again for reasons I do not remember. Anyway, while the other boys went out to play games, I used to stay in and read. In modern terms I was a nerd.

My remembrance of reading was being hunched up in a corner, surrounded by stacks of books. I was always afraid I would run out of something to read. I still get this worry. Also I remember my mother telling me to sit up straight. Evidently I would stop breathing in my absorption and then take a gulp of air. Anyway, the source of all my pleasure in those days was my library books and my pets. The library was indeed a pillar of existence in those early years.

The next library which I remember was at Indiana University, where I went as a college student. Nothing special, except that I used it and became more hooked on books. Then graduate school, and after that out into the real world. I began as a museum curator in anthropology, where I learned the value of reference books. We had our own museum library. After that, a nomadic, research-academic career, using books for preparing professional papers and teaching anthropology students. We used texts mainly, but put books on reserve for extra or special projects. Also I was continually reading anthropology books and articles to keep up with the changes in the field. The library was always there, a comfortable feeling to an idea merchant. And so after a 40-year career as an idea purveyor, I was put out to pasture, retired. It was okay so long as I could get books. And so while still continuing to write, I began to read what I had not had time for during the teaching years. And I continued to write, as I had during most of my life.

One thing led to another and before you could say "gigantopithecus horribilis" I was an independent publisher, purveying my own books. And I found out that the small publisher in particular had to locate his/her niche, which means the kind of people who would buy those particular books. And incredibly satisfying as it was, I found out that one of my special niches was libraries. I had had little idea how all those books got into libraries. But as a publisher I soon found out. There were special book exhibits for librarians all over the country, cooperative mailing schemes to send them flyers describing the books, catalogs to do the same, journals to review them, and special distribution houses through which they would order. It was especially satisfying for me to have my first book reviewed in the <u>Library Journal</u>, the premier national magazine for librarians. My other main group of purchasers were college teachers, choosing my books for class use.

And then I "discovered" my local public library. It is the main library in North County of San Diego and is probably the second biggest, and busiest, in the county, the main one being in the city of San Diego. What a distance in time, distance, and usage from that little brick building in Indianapolis. But still it is of the same family and I shall regard them both fondly for the rest of my life.

I look with even more awe now at the Escondido library than I did at the Indianapolis one. Of course in Indianapolis I was just a small boy from the working class and the library was the first institution I knew for expanding my world view. But when I got to the Escondido Public Library, expansion of my vision through books had become a habit.

Moreover, after those sixty years of using books, I had also learned what a powerful institution the library was. And though I am speaking of all sizes now, I particularly mean the local public library. It is the democratic institution "par excellence," in significance far beyond the political party or the court system. This is the place where you can learn any of the ideas that humankind has developed.

The Escondido Public Library is not imposing as a building, a modern slab construction. It is easy to get to, on a through street, perhaps five minutes from the freeway. And not being in a metropolis, there is adequate parking. When I go there, I park just around the corner from the entrance at a curb where there are always several non-metered, non-timed parking places. There are a few benches and a coffee bar in the covered walkway leading to the main doors. There are usually small groups or individuals in this area, and usually doing something with books.

It is when you enter the automatic door that you see what a library can stimulate. There are people everywhere, not in crowds but spaced frequently. And almost all are doing something. It is a busy place. These people have goals. Just inside the entrance are three check-out cubicles. The checkout girls are stamping the books electronically before the people can take their treasures out the other side of the main door. There is practically no waiting to get books marked, which as I said before, is typical of the place. Unlike banks or most post offices, the library provides service.

Then straight ahead is a block of computers where people are searching for specific books. With that typical computer gaze, they stare intently at the screen. If they find what they want, pencils and note paper are handy. The computer system set up by the library is simple and can be learned in a few minutes. But it doesn't matter, because any librarian, of which there are plenty, will help show anyone who never got into computers. This is true of any operation in the library. There is always someone to help. I compare the library with big discount and department stores. Because they are trying to sell their items, one would think they would make it easy for the customer. But unfortunately this is frequently not the case. I avoid going to them unless I know exactly what I want. Advisers are too hard to locate. Anyway, to continue the survey, just beyond the computer bank is the reference counter. There are one or two people on hand there also, and they are knowledgeable. I have gone to them with many requests about the publishing business and have always got information in about five minutes. And here also there is very little waiting. I can remember no time when there was more than one person waiting.

Against the right side of the large single room are a variety of special counters, one for reserving books, one for getting library cards or to pay fines, one for video or other tape loans, one for books in braille, and I believe a couple of others. Straight in the back is a used book store where duplicates or unwanted books are sold for very reasonable prices, most between 35¢ and $1. I frequently stop there to pick up one or so for my casual reading. Then when I am finished with them, I simply bring them back as a donation, whereupon the library can sell them again. For a person who grew up during the

Depression years, I find this recycling scheme very good.

Off to the left is a section for journals and magazines, and cubicles and tables to sit at for reading them. There are usually quite a few people at these places, reading and taking notes.

At the front left is a place of copying machines. These are do-it-yourself for 10 cents per copy. I have dropped many a dollar into these machines, copying lists for mailing.

The reader of this piece may ask "But where are the books? Are there no stacks?"

Sure there are, probably 2/3rds of the bottom floor on the left side. It isn't as imposing as one might imagine, and for one reason. Branch libraries now depend heavily on loan services. They have had them for as long as I remember but now it is a mainstay. The net for this library covers the county as a minimum, though I would guess it is larger. I have another old friend, one I had not seen for 45 years, who tracked me down as an author. He was living in Minnesota and got one of my books through library loan from Illinois. In Escondido I can find a book on the computer, track it down to whatever library it is at, in the county or farther, and have it loaned to me from my library. I have done this several times, and in each instance it has only taken a week or so. Not a bad system, eh?

There is a second floor also which has a large children's section and the administrative offices.

So there you are, one of the greatest technical inventions of the civilized world, the book, housed in one of society's most efficient social inventions, the library. It makes one almost glad to have been born in the democratic era. I think on the next revision of my will I will put in an item dedicating part of my inheritance to a library. While I am no Andrew Carnegie, I will be able to offer something significant, and certainly the will is there. And in the meantime I will continue using the library for flipping ideas. Hopefully, I will also continue selling them books.

8

From the Top of a Hill

I live in a house on top of a hill. It is special in many ways. It is very private and encourages meditation. Also it has a spectacular view of Southern California hill country. And far from last, it is the house of my recuperation. I call it simply Hilltop.

This is the first time I have lived on a hilltop. When I was a boy, I spent long vacations at the house of my grandparents, who lived in a house part way up a hill in Indiana. I spent some of the happiest times of my life there, catching lizards on rail fences and frogs in marshy places while wandering through the woods and fields. The memory of that hilly place has remained with me all the following years and perhaps helped cause me to seek Hilltop.

Most people live on flat land. Except for one other house, in the suburbs, which was on a slope, all the other places where I lived were on flat land. There is a good reason for this. If they are not built on solid foundations, houses on hillsides can slide down. This is a serious problem in California, because some well-to-do people do build houses on hills. They like the view of the ocean or mountains. But most houses in California, and in the rest of the world, are on flat places because these are normally better for the everyday activities of life.

My situation is different. I came to Hilltop to be an avocado farmer as well as to think, write, grow vegetables and fruit and watch the wild life. Avocado trees do very well on hillsides in this region. They do not slide down slopes because they have wide-spreading roots which probe into the soil. Also avocado trees like to be well-drained. On hillsides water does not stand in puddles. Avocado cultivators hereabouts say that their trees do not like to have wet feet. And finally, the hillsides in this region, northern San Diego County, are better for sub-tropical plants, like avocados, because there is less frost there. Cold air descends to the bottoms of the canyons. On cold mornings in December and January heavy coatings of rime frost can be found on the grass at the bottom of the hills and often enough hard ice in puddles. The leaves of trees on the upper slopes will be covered with droplets of water only. The valley bottoms are fine for the California live oak but not for subtropical aliens like the avocado.

I was never anything more than a part-time farmer. My neighbors and I are mainly city people who bought small avocado farms for investment or retirement. Mine serves both functions.

I believe having a small farm fulfilled a lifelong dream for me. When I was a small boy, people used to ask me what I wanted to be when I grew up. For years I would say, "I want to be a farmer."

The only kind of farmers I knew then were my grandparents and other relatives in Indiana. They cultivated corn and wheat and raised cattle and pigs and chickens. It was called mixed farming and is almost completely gone now.

I liked being around all those animals and plants. But I wouldn't want to be that kind of farmer now. My relatives, like most other American farmers of the time, would kill their animals and eat them. Later they and their descendants learned to buy their meat from supermarkets. And even though I did it myself sometimes in my earlier years, I never did like to kill animals or have them killed so I could eat them. And so about fifteen years ago I stopped eating animal meat. I continue to eat fish, eggs and cheese. I do not think it is wrong to eat meat. And I know that man has been a flesh eater since the dawn of history. However, I feel better since I stopped eating animal and bird flesh. And as an added bonus, during my lifetime the medicos have given more and more praise to vegetables and fruit for better health.

I also got Hilltop because I like to grow things. It is almost a religious experience for me to care for a seed while it turns into a seedling, a plantling, and finally a productive flowering and fruiting plant. The wide variety of plants at Hilltop helped satisfy that need. This is the first place I have lived that I had enough space to plant anything I liked, including many plants and trees which I did not even know existed before Hilltop. Southern California is particularly blessed in its climate. It allows the growing of tropical bananas while also permitting temperate zone fruits like apples. All these plants make me feel like a rich man more than any amount of money could. I often see Hilltop as a church where I am worshiping at a great green altar. If I had to invent a god, he/she would be one of life and growth, perhaps of the sun or chlorophyll. The main offering would be plant products.

I like to write and for that I need privacy. Hilltop is very private. Though there are neighbors on my hillside and across the canyon, they are not close. Hilltop is not like a place in the city where people passing one's house can look in. My house is not closed to public eyes as in the city, the details of which I will describe in the next chapter, "Curtains".

But there is another kind of privacy at Hilltop. The sounds of people are heard only occasionally. There are no sounds of sirens, whistles, or screeching brakes. Those are city sounds. The occasional sounds of people come from cars passing on the access road, normally in the morning and evening. By standing outside and listening intently, the distant hum of car tires on the freeway a mile away, can be heard. But this is a faraway hum, no more intrusive than the buzzing of bees. The freeway hum cannot be heard from inside the house. Occasionally a small plane or helicopter passes on its way to spray an avocado grove. Hilltop is not on the flight path of any big planes. Every few weeks there is the distant "whomp, whomp" of big guns being fired at Camp Pendleton, twenty miles to the west. But those sounds hardly seem to be coming from the same world.

It took several months at Hilltop before I realized another city sound was missing. Unless someone else is at the place, one rarely hears the human voice. The Guatemalans in the trailer at the other end of Hilltop are seldom heard. And there are no quarreling neighbors or children calling to one another.

Twice in the dozen years I have been here, there have been

families with teen age boys. And as they do in late 20th century America, they have raced up and down the road for a few years -- until they got their first car. In my day we went from bicycle to car, but we were working class. And then too all the variety of big wheelers had not yet been invented. But in the total rhythm of Hilltop, the off-road teen-agers have been only brief interruptions.

The outdoor sounds at Hilltop come primarily from birds, increasing in volume through the breeding season. If one walks about outside, one hears the occasional scream of a cruising hawk or the chattering of a ground squirrel which has spotted me or the flying hunter. Dogs bark sometimes, both mine and those of the neighbors. As dusk settles, the coyotes gather in the canyons to set up their yip-yowl serenades.

Thus Hilltop is a fine place for thinking and writing. I have done both in many other places, but always with a greater degree of effort to block out alien sights and sounds. For someone who has had to struggle for a lifetime to achieve the self-discipline necessary to write, it has been comfortable finally to be somewhere where the distractions were few.

But not only is Hilltop private, it has some of the most spectacular scenery I have seen. Through the glass doors and windows one sees sky, leafy green walls, and off in the distance, rolling hills. There are groves of avocado or citrus trees interspersed with native brush. This is the heart of California's avocado belt, halfway between Escondido and Fallbrook. The hills go on as far as the eye can see. Off to the northeast is the long ridge of Palomar. It is the highest land in sight. The realtor's description of Hilltop included the item, "280 degree view." That is true, but I think it is the understatement of the year.

I like to write but it is hard for me to string words together for long stretches of time. After two or three pages, my thoughts come slower and I tire staring at the typewriter, now computer screen. This is a good time for me to get up and walk around. The walking shakes loose the word-jam and the judicious phrase comes to mind. During writing stints in the past, I walked with little more than the walls of rooms to stare at. And even where I had a window to gaze from, the vista was always limited. What a difference at Hilltop. Now from just outside the house I have mixed chaparral-avocado land extending off until it merges into the oak-pine highland of Cleveland National Forest. And all that I see is topped by blue sky in clear air. What jaded writer could ask for more.

A bonus I had not expected was the depth of view from Hilltop. It is much more than a mere physical perception. Rather I have found it to be a mind expanding experience. Men have sought different ways to get beyond the ordinary, to enable the mind to soar. Some of the ways have been meditation, prayer, body control or stimulation. One of the mind expanding experiences I have found is bicycling. My thoughts have often soared while I have been pumping rhythmically across the countryside on my ten-speed. And now for the first time of my life being at a particular place does the same thing. Since coming to Hilltop, I have written about all kinds of topics I never would have if I had remained the pure urbanite I used to be.

Like so many Americans of the 20th century, I have drifted back and forth over the face of the land in my lifetime. The kind of stability that once existed in rural America has not been part of my heritage. I have gone from one place to another, primarily for better jobs. Coming to California was no exception. I was offered a good academic position in Los Angeles.

I had longed for some kind of permanence for many years. At the last two places before Hilltop, I thought I might put down roots. But it was not to be. However, I feel now that I have more chance for permanence than ever before. And it is a comfortable feeling.

I did not come to Hilltop primarily for monetary gain, though this has happened. Rather I came mostly for therapy. I could properly call Hilltop my house of recuperation. There are various ways of coping with life's problems. Some people seek help from friends, relatives or religious leaders. Others go to doctors or psychotherapists. A few try to solve their own problems unaided. I have found that writing works well for me. I write my problems out of my system.

A not uncommon way to try to improve conditions is to go away. One can flee the scene of one's disappointments, at least temporarily. I have used that method. I recently read an excellent autobiography, Blue Highways by William Least Moon. He was trying to deal with a separation from his wife and a loss of his job. His solution was to travel the side roads of America and write about it. When I came to North County, I too had gone through some unpleasant occurrences, a serious illness and a bitter divorce. I ran away for five months, including out of the country. But when the legal dust had settled and the property settlement was done, I was again rootless. I decided to get a place where I could write while getting as much from daily life as possible. Also there was a tax incentive. After the divorce, my take-home pay became considerably higher. And I had heard that avocado groves were excellent for keeping more of one's income.

My wounds were only partially healed when I came to Hilltop. But the illnesses had been the most serious of my life. However, as I became more deeply immersed in the activities of Hilltop, those dreary times receded further and further into the background. I became too busy arranging, planning, cleaning up, decorating, planting and trimming to think much of the unhappy period before. The sights and sounds of Hilltop, always fascinating, rapidly replaced the sad thoughts of hospitalization and the weakness of post-illness, the unpleasant conversations with divorce lawyers and how affection had turned into hate. I watched the continuing panorama from Hilltop, the early morning flight of crows, the roadrunner bobbing past my window, a lizard in its beak, the April rain clouds sheeting across the hills, the early morning fog enshrouding the hills two-thirds of the way up like a Chinese landscape painting, the gopher snakes making love in my garage, and on most days the glorious sun pouring life onto the land.

In those early days, I was teaching full-time in Los Angeles. With a four- day schedule, I had planned to stay in town three nights each week. However, I usually had one day teaching, then one day off. And though it was a hundred mile drive, I could not resist returning the night before the free day. I would come down the freeway wearily sometime between ten and eleven at night. Driving up my hill, I would stop just in front of the garage door. My Jack Russell Terriers would come bounding out, joyous to see me back. I would try to pat them evenly while they jostled one another jealously. Then we would all troop into the house. I would quickly get my long-beam flashlight and hurry outside to scan everything I could. What had grown, where had new flowers blossomed, what plants had set new fruit, had the gophers done any damage, had the swallows returned from the tropics to refurbish their nests of last year? Then I would turn off the light and stare up at the sky dome, the stars wondrously numerous in the clear air.

I would walk over to the edge of my little spread where the avocado trees end and the brush land begins. A path is practically always visible, no matter what time of night one goes into the chaparral there. I would walk out to a little cleared space and gaze across the canyon at the pinpricks of light that marked distant houses. My musings would be broken by the distant tuning for a coyote serenade. A few yips, then howls until the group got into whatever tune is appropriate for them. Then the canyon would echo with their music and I would feel a deep kinship for those noisy survivors. I would turn back, deeply content, saying audibly, "Have a good night, brother."

9

Curtains

Anthropologists have become concerned with social space, a field of study they call proxemics. It seems that all social groups have evolved rules to govern the amount of space permitted between individuals and objects. But there are not only rules to control how close people and things can be to one another, there are also rules to control how much and what one is permitted to see in different circumstances. There is public and private space, the first more restrictive than the second. A house is governed by rules of space. Nudity is permitted only by certain persons in certain places. In particular, window coverings are used to close off private space, especially by females. My house at Hilltop is an exception because it was not furnished by a female.

Why do we put windows in houses? That seems easy. So we can look out. Maybe also so we can let some light in. Both explanations are true, but only partially. Because if windows existed solely so we could see the world outside, then why all the coverings, all the various kinds of curtains? I know that in all the houses I ever lived in, except one, that window coverings were the first order of business. The one exception has been the house where I now live in San Diego's North County. Originally I put drapes only on the south and east side at Hilltop, and their primary function was to keep the heat of the summer sun from building up inside. Now there are vertical blinds instead of drapes on these windows.

There are two more windows with coverings, one to close off the living room from outsiders and one on the bedroom window, facing the avocado grove. Only rarely do outsiders come by the bedroom window, but this room is especially supposed to be private. Oh yes, there is also a cafe curtain over the kitchen window which blocks off any eye-view from the road. These latter additions came with Jessica who felt uncomfortable at being exposed. She had curtains or blinds over every window in our Culver City house.

There is public and private space and rules to enforce them. Public space is more restricted than private space. One cannot take off one's clothing outside but one can inside. The front yard is public, the back yard iffy.

I once knew a lady who considered herself quite liberated and who had a swimming pool in her back yard, enclosed by a fairly high wall. Without a ladder, you could not look into the pool area. She used it occasionally for sun bathing and would sometimes take off her halter. Once when I came to visit she was furious because workmen on a nearby roof had been ogling her. I asked her, " Did you expect them to turn away?'

She, "'Of course, I was in my own yard." She was only partly right. The workmen had to be on the roof they were repairing, and how many men in 20th century America would refuse to look at a lady's breasts? It was an unusual set of circumstances when cultural rules were not serving their usual function.

The same kind of problem occurred in our Culver City house. Jessica thought of the back yard as private space. So not being a dedicated cross-culturist, when she spotted the Mexican gardener urinating in one corner of the yard, she tried to get him to stop by calling out to him. It was a cross-cultural conflict since male Mexicans of the working class feel comfortable urinating in public space and are not prosecuted for doing so. Women are supposed to turn away if they do not want to be embarrassed. The gardener had to know about the difference in Anglo-land or he would not have to tried to hide. But still he evidently felt it was alright enough for him not to have to look up a men's toilet and break up his work schedule.

Jessica extended the private space of our house to include the neighbor's yard. A fairly high wall separated our yard from that of our neighbor, Harold. Also he had made it even higher by putting up a wood lattice addition. However, there was still one particular place and angle through which one could see his back yard. And though it was on the opposite side of the yard, Jessica claimed she saw Harold doing chin-ups without any clothes on. Harold had a spa in his back yard and he claimed he did the chin-ups after getting out of the spa at the time in question. He used to bathe without swimming trunks, as quite a few males or couples do. That is culturally okay since no-one else is around to see.

Was visual space being violated according to our cultural rules? Both parties considered their point of view correct, a not unusual occurrence. But this is again a grey area in which the standard cultural rules do not work perfectly. I am not sure what a judge would rule.

Some leeway is permitted according to the particular social group, exact location or occasion. People are allowed to expose more of their body on a public beach than downtown. In fact there are even a few nude beaches, particularly in southern California. But they remain oddities in North America. I have been at Black's Beach, one of the best known locally, and I found that people were clustered around the few nudes, many taking pictures. Europeans and Australians are freer with nudity on beaches, though they will not permit it in town. Before Christian missionaries got to them, there were natives in some parts of the world who wore nothing on or off the beach.

Social protest can be made by exposing a part of the body improperly. There used to be a practice of "mooning" in which young people would expose their buttocks from the window of a moving car. Also there was a period when nude young men would streak through a crowd. I remember a young man waving his penis at a passing train south of Santa Barbara. He could afford to take that chance on the probability that the train would not stop to let law enforcement officers off. All these instances were young people protesting the strictures of society, a common occurrence in the 60's and 70's.

What is proper to see is especially controlled in public places. People have to keep their bodies covered properly. Many restaurants refuse to serve people who wear the wrong costume. Special facilities are provided for private activities in public buildings. Cubicles exist for people to go to the toilet because they have to expose private body parts, men urinating excepted. But even for males urinating, there are informal rules for how much one can look in men's toilets, as my colleague, Elliott Oring, described.

Of course, urinating outside in Anglo-American culture is a legal offense. An interesting occurrence was described in the film "Harry and Tonto" of improper peeing in an outside public place. After a long night of gambling in Las Vegas, and after looking unsuccessfully for a toilet, the old man, Harry, went outside and urinated into a potted palm. And even though he shielded himself from public view, his body position was enough to give him away. A squad car rolled up and Harry was carted off to jail.

In hospitals special curtains on frames are provided to keep patients from being exposed or for hiding medical activities. On the other hand, hospitals take certain liberties like dressing patients in shortie gowns open at the back for the convenience of medical personnel. Anyway, sick people are generally permitted more leeway in behavior in most cultures.

As many of us know, there are different customs in other cultures as to what is proper. Frequently a difference causes a shock. I remember the first time I took my wife to France and she came rushing out of a toilet in the Paris Airport, surprise on her face. "There were men in there!"

I couldn't keep from laughing. "I should have told you," I said."Both sexes share the toilets in many places in France. It doesn't mean anything, although the old British belief that all Frenchmen were immoral may derive at least partly from this custom. However, they don't think of it as improper. And I can assure you that no-one would have bothered you. And you don't have to look at the men standing at the urinals since there are closed cubicles for women, aren't there?"

But the really private place in our lives is our home. It is not just where we hang our hat. It is the place where we can do the most outlandish things, sometimes in the open, at other times in closed rooms. It is the reason we say "A man's home is his castle." The home is the place where we no longer have to use a public cover-up. But that only works if we make sure we are separated from the public domain. And so the openings that were first designed to bring the outside in and let the inside out, windows, have to be covered. Wide open windows would be fine if there wasn't anyone to look in.

But that certainly isn't the case in the cities where most Americans live. Houses are built on 50' X 150' lots, and they face a front street where people walk and drive by, not to mention salesmen and delivery people who come up the front walk. Sometimes the back is cut off from public stare. But the sides are frequently accessible to curious eyes. The solution is window coverings, hopefully ones that will let light in while keeping eyes out, but which can be opened when prying eyes are no problem.

Although curtains, shades, etc. are put in windows all over the house, the place of greatest concern is where the body is exposed most. The bathroom usually does not have any sizeable windows, but bedrooms frequently do. I learned this when I was a young boy. My friends introduced me to "peeping." They had found certain bedrooms in the neighborhood where the windows had a crack at the side when the shade was drawn. We would creep up and try to peer through. When I grew up and got married, my wife was always very careful about having the shades or blinds drawn in the bedroom as soon as it got dark. She insisted on this before a light could be turned on.

So now my house on the hill has caused me to think specifically about public and private visual propriety. There are no prying eyes, which makes a big difference. The place is so located that people do not pass by often. Then too, the house has been furnished by me, mainly without any woman's help. It is a man's house in the country. And I think there is no doubt that men and women have separate expectations as to what is proper to be able to see.

Hilltop is a mile from the freeway on an access road where most of the people who walk by are Latinos on their way to, or looking for, work. They are circumspect and their presence is signaled by noisy dogs. Also the work seekers keep thirty or forty feet away from the house, waiting for the "patron" to come out. There are also two children up the road who pass twice on weekdays, going and coming from school. Finally, there is an older couple, the Baumanns, who come by most days. This is their constitutional and they only miss when they are not feeling well.

Most of the local residents come by in cars, but not frequently. Once in a while a UPS or garbage truck comes by, but otherwise people rarely come unless there has been a pre-arrangement. No delivery or sales people come this far. It is not a through road. Except for one side of the kitchen, the house faces citrus and avocado groves, the chaparral beyond. No outsiders come from these directions.

Domingo and his wife, Katerina, lived in a trailer at the corner of the avocado grove for a few years, but they kept to themselves generally except for the one day a week that Domingo helped with the chores.

This is a kind of privacy I have not experienced in my seventy odd years. The only place I can think of that was at all similar was my grandparent's farm in Indiana. Their house was on an isolated, unpaved road, with the nearest neighbor across on the opposite hillside, about as far as the houses across the canyon. Also, many people did not wander by there either and the few who did were always signalled by barking dogs. This kind of privacy must have been usual in 19th century America and only continues in a few rural pockets.

With this degree of privacy, public and private customs become similar. What one can only do in a bathroom or bedroom in the city, one can do outside here without violating any customs or breaking any laws. I had no problem urinating privately outside when Jessica was not here. When she was, she tried to impose city rules. The ultimate in freedom was the instance several years ago when I had sexual intercourse with a lady under a tree on the hillside (See "Love Affair with Avocados). That was only the second time I had done that outside. The other time had been about 50 years before when I was a soldier in the French countryside (See "On Being a Latter Day Veteran"). I am sure such an occurrence fulfills a special fantasy.

Customs change as do all things, but not overnight. People do not immediately change their ways to fit new conditions. The renters in this house before me, like the residents of the area generally, were city people and had city customs. Moreover, they were a family group and the woman's viewpoint prevailed. So when I bought Hilltop, there were curtains or drapes on every window. I am sure the ladies had put those up as they had done in the city houses where they had lived before. They had assumed there would be prying eyes.

I was impatient to bring the outside in. So I pulled the drapes and curtains to the sides and rolled up the shades. Then piecemeal, I removed one after another, hauling them off to the county dump. And apart from the drapes to keep out the summer sun, for several years I had no curtains. Now there are vertical blinds partly because the drapes wore out and partly because there was a feminine period in the house.

However, I still find it very pleasurable to roll over, facing east to greet the rising sun. It is one of my many rituals as a sun worshipper. The north- facing sliding doors of my study go undraped. After all, it faces the green shield of avocado trees ten to twenty feet away and strangers never pass by. The bedroom window on the same side does have blinds, but bedrooms are special, particularly to women.

10

On Sorting Garbage

A few years after I had got a divorce and bought Hilltop, I went to the dictionary to sort out some bothering questions. What was a husband? Was I a husband? I felt like I was, though the state said I was not. She had sued for divorce, not I. Or, did I practice husbandry? If so, how without a wife? The American Heritage Dictionary said the following: Husbandry, obs. 1. The care of a household; 2.The control or judicious use of resources, conservation; 3. Cultivation or production of plants and animals. "To husband" was defined as "to manage prudently and economically" or "to use sparingly." The noun, husband, could be a married man, steward or frugal manager. I decided then that I was a husband, no matter what the court said. How did I do that?

As usual with many of us loquacious academics, I will go off on a tangent. The primary meaning of husband and its derivatives obviously comes from the age of agriculture. It derives from the second great revolution in human knowledge, the domestication of plants and animals. The first basic achievement of mankind was the invention of tools as a primary way of adapting to and exploiting the environment. And though tools made man a more efficient gatherer, he was still only doing what almost all the other animals were doing, collecting wild products. But when much later in his evolution, he learned how to control the lifeway of some of those plants and animals, he was taking off on a wholly different path. And apart from a steady improvement in technology, man still has not learned a better way to get food and other necessities. He is still a domesticator, no matter the amount of machinery he uses.

The ramifications of the revolutionary new way to exploit the environment have been enormous. In particular, it enabled many more people to be fed and thus caused a great population increase. If one thinks of more as being better, this was good. If one is thinking of the quality of life, it is not so simple. Agriculture could have made life easier for individuals if their social groups had not continually expanded; but for various reasons, in one part of the world after another populations got too big to feed everyone adequately, no matter that overall agriculture was much more efficient than hunting and gathering. The increased population was eating up the surplus. Traditional agricultural societies have almost always had big families, while hunter-gatherers and urban man have had fewer children.

There is little doubt that gathering peoples had hard times, as did their animal counterparts. There was starvation on occasion, when the wild plants failed or the animal herds did not arrive. However, the really disastrous famines of history have been in dense populations of farmers. The main problem is that agricultural peoples have had access to more kinds of goods than gatherers and their needs increased. They had to work more to get more. What cultivator was willing to get by with a wooden digging stick once the plow was invented? And who would be willing to hunt with a bow and arrow once the firearm became available? Even when they suffered from the change, tribesmen ordinarily wanted new technology. Guns from traders turned out to be disastrous to North American Indians. They eagerly used the new devices for getting fur-bearing animals for Euroman. And when he no longer wanted fur, the tribesmen had so geared their lives to trading post products that they couldn't go back to the old ways of hunting and gathering for subsistance. There is good evidence now that true leisure went with the hunting and gathering way of life rather than with that of industrial man.

An overall consequence of greater numbers of people and improved tools has been the ever-increasing ability to ravage the environment. It has become fashionable in recent years to give tribal peoples credit for treating the earth much more considerately than industrial man. And though the tribal man, who lives closer to his environment, is more likely to include it in his religion and philosophy, the real fact of the matter is that tribals did not have the numbers to cause much destruction. They undoubtedly threw stuff they didn't want into their rivers, but how could they foul these up like industrial man could with his machinery, chemicals and much greater production? The hunter and gatherer was also nomadic, so he left behind what little trash he had to throw away. The gnawers of the world did the rest. The campsite of the hunter-gatherer is a rare kind of site in archeology. However, when man settled down in a fixed place in the early cultivating societies, he started leaving trash piles behind. The Pueblo cultivators of the American southwest used to toss their broken pots, corn cobs and rabbit bones over the edges of their little mesas. And in the eastern U.S. broken pots became the diagnostic items of archeology for village cultures, as stone tools had been for hunting-gathering ones. Without trash heaps, containing stone tools and potsherds, the science of archeology could hardly exist.

But with small populations, even of simple cultivators, the need to be parsimonious in the use of resources was less important. Even so, some major events of what we would today include in the term, ecology, probably took place. There is a hypothesis that the cold weather mammals (hairy mammoth, rhino, wild horse, ground sloth) were at least helped along to extinction by early hunters. And there is also the idea that the prairies of North America were at least helped along to be grasslands by the fires of the Indians. But still ecological destruction, like the dust storms of America in the 1930s, or massive pollution of the air and water, were beyond the capacities of these small groups with primitive technologies.

Besides a big increase in population in the agricultural age, there were a number of technological improvements that brought significant new ecological problems. One was the invention of metallurgy, and particularly of iron, and even more particularly, the invention of the steel axe. Hunting-gathering peoples, and even simple cultivators, got by with implements of wood or stone. And though these gave man as a species an important edge in evolutionary competition, they just were not complex enough to cause major environmental changes. The steel axe was a wholly new story, enabling <u>Homo sapiens</u> to cut down vast forests. Trees had to go if man was to make it by cultivation, because almost all the food plants needed full sunshine to be productive. And go they did. First it happened in the Middle East, where iron was invented, then in Europe, then all over the world. Now it is going on in the last forest areas, particularly in the Amazon, while the forest destroyer of the last two thousand years, Euroman, makes a hue and cry about the destruction of the last great tropical rain forests. Thus, the steel axe, and later the chain saw, have changed the earth enormously.

But there were many other technological innovations that contributed. Here are a few others. The steel plow was a device that could cut deep into the topsoil, enabling much bigger crops but exposing the land to greater erosion. Then there was sheep, a very productive animal of the era of domestication, giving milk, meat and wool. Not bad, but on top of that, sheep could manage in a semi-arid environment, even on the edge of the desert. That's good for a sheep raiser but may not be so wonderful for the land, in particular because sheep eat the whole plant. They even clip out the roots, in places where plants live precariously.

One other major change contributed to all the rest. This was a change in settlement types, followed by many changes in technology. The greater efficiency of agriculture permitted a new kind of social unit, the city, to emerge. Vast numbers of people could be supported in a very small area. And this enabled more technical change, ultimately the industrial and chemical revolutions. I'm sure I do not need to stress how much pollution these caused. To make a long story short, therefore, the environment came to be at risk. Nowadays, many consider the whole planet threatened.

But quite awhile ago, at least as soon as agricultural man was well settled in Europe, some of the problems were recognized. So in early England men began to practice husbandry, the parsimonious use of resources. They tried to use the new technology without destroying the way of life, to learn to be frugal managers of the domestic plants and animals, as well as the soil, air and water which supported them. And though this happened in many parts of Europe, in many of the new lands taken over by Euroman elsewhere greed prevented judiciousness. Extinction of many animals was the order of the day, beginning with the navigators, who killed off many island species, simply because they were unafraid of man and could be eaten. But when Euroman got established on the new lands, he was even more callous, killing off in mainland North America the passenger pigeon, the native parakeet, the wolf, grizzly bear and buffalo, sometimes to extinction, at other times to the verge of. A farm became a place where after eliminating most of the native trees, one cultivated a number of selected plants and animals, mostly from Europe. Anything that was in the way, including the native peoples, was eliminated. It is not hard to understand why there are so many types on the endangered species list.

And so we come to the age of ecology, when after all the ravishment of the land and its native inhabitants, it became popular to try to save the environment. It was finally recognized that man himself was part of the environment and would last only as long as it did.

One way to help restore the environment was to try to stop further devastation by commercial interests. Many organizations have emerged to prevent killing off this species or that, or to stop cutting down virgin timber, or whatever remained, or to keep factories from dumping their wastes into the air or water.

Okay, but what else? Is there nothing we can do personally? Cannot we as individuals use resources judiciously, can we not become husbandmen again, like our post Iron Age ancestors? I am sure that more can be done than generally is. But it requires an attitude, a general frame of mind -- to use what there is without damaging the environment further. It will not be easy after the decades of wasteful consumption we have been trained to indulge in. But then, no change is easy, though one of the best times is when a crises occurs, which must be now.

This was a long tangent which left out one important matter, the connection between being a husbandman of land and a husband of a wife. I think the connection is fairly clear, though as usual this old professor has a few words to add. A husband of land lived as a part of a family which during that period was patriarchal. One of his prime resources was a wife whom he managed also.

Now of course much has changed, both in agriculture and in the nature of the family. In industrial urban societies agriculture has become a kind of industry in the field, managed by professionals in the employ of corporations. Moreover, the family has lost its patriarchal stamp and no-one is supposed to manage anyone else. It is not exactly like this in all instances, but even children are supposed to do their own thing rather than being managed for the sake of the family. We still have husbands and wives, but I'm afraid their respective roles are a far cry from those of the patriarchal family. So we are left with the possibility only of managing the other resources, the nonhuman ones. And which helps answer the initial question and to go on to the final section of this piece, to wit: I was and am a husbandman, though an on-again-off-again husband.

    I sort garbage. There are three receptacles. One is for burnables, primarily paper products. These are destined for burning in the cast iron heating stove in the living room. I get my groceries in paper bags at the super market for lining my burnable containers and put full bags into the stove. Another container in the kitchen is for non-burnables, primarily the products of industrialization, mostly glass, plastic and metal. Through the years this has been discarded in several ways. In the early days at Hilltop, I had a paid garbage pick-up service which took my sacks of non-burnables to the local landfill. Then for a while I gave the bags to one of my roomers, who dropped them into a dumpster where he worked. Then Jessica and I got a house in Culver City which had a regular garbage pick-up and to which I would take one or two plastic sacks each time I went there. In all instances the non-burnables ended up in landfills. A third container is for rottables, mainly vegetal parts not used for cooking. Since I do not eat meat, I do not have any parts of dead animals to put into the container. At first I emptied it under an orchard tree or on the side of the hill in the leaf mulch of avocado trees It decomposed rapidly. Probably the most visible discards were orange rinds, but even they disappeared in about a week. If meat or bone had been included, rats would be attracted. And though I admire the rat as a powerful competitor in the evolutionary sweepstakes, I do not want to share my living space with him.

    I did not begin this sorting procedure from a specific decision to practice ecology. Rather I think it developed from a combination of causes. First off, I have a continual need for a certain kind of orderliness. I don't like things to be out of place, probably a part of my maternal heritage. Anyway, one of the first things I do each day at

On Sorting Garbage

Hilltop is to walk about, putting things back where they belong. Even a little asymmetry bothers me. A rectangular throw rug not parallel to the nearest wall must be straightened. One of my minor irritations with Jessica was that she put things in piles that did not seem to have order. This bothered me more at the farm than in the city house, I think, because I was more concerned with Hilltop. It isn't that Jessica had no system of her own. She was actually quite good at finding things in her piles, as well as in her incredibly packed purse and toilet kit. But it was her system that didn't make much sense to me. Anyway I have my own way of keeping things in their proper place and I suspect this influenced me to sort garbage. Each kind needs to go to its own proper place.

But I also practice husbandry with the garbage, using what is usable. So the burnables provide heat and ash (for fertilization) while the rottables are for soil enrichment. I don't know how to use the non-burnables and so do what most everyone else does, send them to the landfill. This is a problem which will not go away completely, though it may become less important as we learn to use more burnables and rottables.

Several visitors have commented on my system. Jessica was amused, though she went along. She did know something about sorting garbage since the city of Boston, where she grew up, was doing some version. However, she has come back to the standard American practice of putting all garbage into a single plastic bag. She laughed at me for being a Depression kid and she was probably partly right. I learned during that period, particularly from my parsimonious mother, to try to use everything. Though I can easily afford not to now, I still find it very difficult to throw away left-over food or not to finish what is on my plate. Also I still remember the sight of garbage cans over-flowing with leftover cooked beans across the alley as a desecration. The neighbors were from Appalachia (We called them Kentuckians) but to us German-Americans they were wasteful outsiders.

And I suppose I have absorbed some of the current feeling about ecology. I did not sort garbage before I came to Hilltop, and in all the city places where I lived before it was easier to toss whatever I did not want into a single container.

Anyway, I now try to use everything I can. I keep koi,

decorative Japanese carp, in a large concrete pond just outside the house. The pond was here when I first came. I learned later that it was a back-up reservoir for irrigation, a reservoir pond. But when I came, it had trash in the bottom and shallow water from rains, milling with tadpoles and mosquito larvae. I cleaned it out, did a minimum of patching and had a local fix-it-guy put in a simple filtration system. I filled the pond with water, put in chlorine, and used it for 2-3 years as a swimming pool. I was living alone in those days and did not use the pond very much, so I decided to turn it over to true water creatures, fish. I cleared the chemicals from the water and put in a half dozen koi. After that the project mushroomed. I added koi until there were about 70. Then an improved filtration system and a bridge. I'm sure some of the fish weigh over 20 pounds now. One of the husbandry features that pleased me greatly was that I could empty a good part of the nutrients from the fish water into the avocado grove. The filter had a catch basin which picked up most of the pond dirt and which could be opened up to release a flow of black gunk down the hill. Directly below the filter was the avocado grove, and though the grove was very productive overall, from chemical fertilizers, I told myself that the trees below the fish filter did better.

I also feel good about the water hyacinths as the primary pond cleaners. These are floating tropical plants which were brought to North America from South America. Where released in warm regions of this country, they have become a real pest because of their ability to propagate themselves. In Florida they have filled many canals. However, if contained, they also clear the water. Their long floating roots trap all kinds of particles and they literally choke green algae. The water turns crystal clear. Moreover, their excess numbers can also be thrown down the hill, to add to the avocado mulch.

Which brings me to another bit of husbandry. Like most trees, avocados do well with a leaf mulch on the ground under them. The forests of the world are places where leaves have been raining down since they started growing. This adds organic matter to the soil, as well as keeping in moisture. Trees grow very well with such a ground covering. In fact, it is now known that soil deteriorates very rapidly in the tropics once the trees are cut down. The tropical sun bakes and hardens the soil and it quickly loses whatever nutrients it had. A special kind of clay, laterite, becomes rocklike in its hardness. Most productive plants will not grow in it.

The hill land in southern California was semi-arid scrub before being cleared for planting, while the soil hereabouts is a pebble-like material called decomposed granite. There was originally little topsoil or organic material, though the decomposed granite had plenty of nutrients. So when irrigation systems were put in, the newly planted trees did well. Soon there was a covering of leaves which got thicker each year. And the trees produced more all the time.

The mulch layer tickled my husbandman's fancy when I first saw it but it satisfied another of my characteristics, laziness. I am innately a very lazy person. And considering this quality, I am surprised sometimes how much I do get accomplished. I think the counterbalance is boredom. I get bored easily, which invariably drives me to work, where I manage a while, until the sameness of the new task begins to get boring also. This is true of writing, teaching, reading and farm work. My life strategy then has been to do the maximum I can without getting too bored then shift to another job. So the fact that I did not need to remove the leaves was a blessing. The only time I did mess with the leaf mulch was when I planted additional trees, at which time I took enough from the old trees to help the new ones get established, until they would start dropping their own leaves.

I also packed in plastic bags grass clippings from Culver City after we got that house. A gardener cut the grass there and I trained him to put the clippings into a plastic bag which I would leave outside on cutting day. Then I would put the clippings around newly planted shrubs on the farm. I doubt that this amount of mulch makes much difference, but it gave me a feeling of satisfaction that I was acting like a true husbandman. I recently read of a lady who was living in Los Feliz, a neighborhood of Los Angeles, and who put all her vegetal discards onto her compost pile in her city lot. Compost piles are an old husbandry practice, though they do require some extra work. My laziness kept me from setting one up. However, I rationalized that my whole farm was one giant compost pile. I would walk out under the trees sometimes, my feet in 6-8 inches of leaf mulch and listen. I could imagine that I was hearing millions of gnawers, converting the leaves into prime soil conditioner.

Then there is the wood-burning stove. When I first came to Hilltop, the renters were using the central fireplace and electric wall heaters during the winter. They complained that the electric bill was very high, while the house was always cold. The first year I had a wood burning stove put in, using the fireplace flue. There was plenty of wood from avocado tree trimmings. Avocado trees grow by spreading and covering up many of the lower limbs. Once a limb no longer receives sunlight, it ceases to produce fruit. At Hilltop these went the way of the chain saw.

It will be apparent to the reader that my kind of ecology is not anti-technological. If it was, I would be denying one of the main premises of anthropology, that the only clear-cut progress we can see in the human way is ever-improving tools. Also there is my natural laziness. I use machines that will do jobs easier. And as is typical of humankind, I have a rational explanation.

Technological improvements, from the simplest to the most complex, can cause destruction. The chipped stone spear point may have been the principal tool which wiped out the big Pleistocene animals, while the iron axe has been the main means of destroying forests. On the other hand, though earth-moving machinery has been used extensively by developers to ravage the landscape, it can also be used for creative landscaping. It is lack of social control over machinery that creates the havoc, not the machinery itself. Anyway, I got started burning locally produced firewood the first year and have continued ever since.

Some dyed-in-the wool ecologists may object to the wood smoke. Firewood does produce smoke, and where there are dense populations this can create smog. I remember cities in North India that were as choked in winter as Los Angeles from fires of wood and coal dust, mixed with cow manure. If everyone heated their houses in Los Angeles with wood fires, the smog situation would be worse than it is. However, Hilltop is in San Diego's North County, still with a dispersed population. It is still the approved policy to burn tree trimmings in the open if fire conditions are safe. So far, there haven't been enough people to pollute the air with cars, smokestacks or wood burning stoves. This will probably change in the next few decades and wood burning, as well as many other practices, will be stopped.

Simply switching to the wood-burning stove helped make the

house warmer. But a year later I also put insulation into the attic, after which the house became very toasty in winter. This has warmed the cockles of this husbandman's heart also.

    Then there are the chickens. The flock simply grew until it became a husbandry fixture. I can't remember why I started exactly, though I suspect it harks back to my boyhood. Perhaps also it is because I have long disliked the modern battery chicken system in which the poor bird has no more space than is necessary for eating and dropping eggs. Most profitable but sad to contemplate. In any event about three years ago a long separated brother-in-law came to visit, and like some other city refugees, he got enthusiastic about helping on farm projects. We decided to build a chicken run and started together. But he left before it was done so I finished it. Then I had to get some chickens. I started modestly with six Rhode Island Reds, which I remembered from my boyhood. They did well and we got two additions, Easter chicks that had grown up. Their city owners were trying to get a permanent home for them. Though most Americans eat chicken with no compunction, most will not eat creatures thought of as pets. Besides, American city people have no experience killing anything for food. Chicken, like all other forms of meat, comes in plastic wrappings from the super market. So Easter chicks are a problem when they grow up. One of the newcomers was an Araucan, a chicken that lays green eggs. All went well and when the chicks grew up, they began laying eggs. Their only fault was that some of them turned into egg cannibals. I noticed that Flo, the Araucan (we named her after my mother-in-law), did not eat eggs. Anyway, that flock was wiped out by a bobcat after a couple of years. The chickens were inside a cyclone fence, but this was nothing to that climbing marauder, particularly since there were many overhanging trees.

    The next spring I decided to try again, but this time keeping the chickens entirely enclosed. This worked and there has been a small flock ever since. We liked the fresh eggs, in no small measure because the chickens were not raised in commercial wire prisons, with barely just standing room, and with their beaks clipped off so they couldn't pick the eggs. Others have told us that the eggs taste much different because of their freshness. Of course, the green shells always make an impression. The yolks and whites are the usual colors. However, the chickens also serve as consumers of food leftovers and garden trimmings. They will eat anything a human will, plus a lot a human will not. Having no cultural taboos, they delight in bugs and worms of all kinds. And then they leave their droppings, which of course makes good fertilizer. I suspect the husbandman in me appreciates this last function more than the egg production.

    I like to cook and moreover I get a great satisfaction from using leftovers. It may well have begun in the Depression when I learned from my mother to use everything. But the tendency was reinforced when I divorced twenty-five years ago. I took over the kitchen duties, preferring to cook my own rather than eat prepared food. Another change in my life then was a great increase in social activities. My home became the place for many big parties. Guests brought food in considerable quantities and much was left over. Instead of throwing them out, I wrapped the left overs and put them in the refrigerator. I became a specialist in casseroles and soups, combining the leftovers in tasty combinations. My dedication to hot chile did not hurt. From this experience I got a wholly new respect for French rural cookery in the era before supermarket packaging. Then when I came to Hilltop I began again where I had left off ten years before. The leftovers from two or three newly prepared dishes became a casserole, and after being warmed over and eaten 2-3 times, it was turned into a soup. The taste got better the more times it was heated up and recombined.

    I also tried to use everything from my prolific garden, if not for me or Jessica, or friends, then for the chickens. When greens were in season, and we could not eat all of them, I boiled them down to make soup stock. This was particularly useful for us because the big commercial companies no longer sold vegetal stock, only chicken and beef cubes.

    I am not truly a squirrel. I like to get rid of stuff I am not using or cannot imagine a use for. However, I do keep things, stored methodically, if I can imagine a use for them. And so I have a shelf of glass and plastic jars in which to keep leftovers or smaller portions. I keep all sizes and shapes to serve for different amounts. Jars are of course one of the big consumer throwaways in our culture, most ending up in landfills. Perhaps someday when we really get serious about recycling, we will stop this waste also.

Hilltop is my ecological corner of the world. However, as a recycling and parsimonious Depression kid, I am far from perfect. I'm too lazy to do everything that ought to be done and perhaps also I'm not imaginative enough. Still, in my own way, I take some steps toward being a husbandman. And I have an attitude which I believe will have to become universal, either through law or voluntarily, and especially in the cultures of affluence, if our species is to survive.

11

The Non-Endangered Species

One of the old masters in anthropology, Leslie White, presented an idea which he called "the anthropomorphic illusion." White claimed that man's belief that he controlled the destiny of his way of life was illusory. Further, he claimed that culture, or "man's way" had its own rules and that there was little in it that man could control. I suspect that White overstated his case, though I also believe he was mainly right. The major tendencies of cultures or civilizations are beyond the individual's control. Cultures come, stay for a while, then fade away, no matter that the members invariably hope and try to keep them going.

And so I am going to extend Leslie's idea in claiming that not only does man have an anthropocentric illusion about culture, he also has one about other species. That is, in the same sense that man believes and acts as if he controls his culture, he also believes and acts as if he controls life forms generally. Like other ideas about the world that have turned out to be false, this one has a certain plausibility. After all, man can legitimately claim that he is the dominant species. He has been able to control, even to the point of extinguishing them, many other species. He has been able to create an endangered species list. I suspect the knowledge of this power gives many men a kind of pride. Thus, for the same creatures that he has endangered, man has set up special provisions to save at least a few of them. It seems that he has been in control in wiping them out and in saving them. So what is the hitch?

For some time now, I have come to the realization that though there are endangered species, there are also some which are doing very well, some which man would be pleased to do without, but whose destiny is not in his hands. In my introductory anthropology classes I suggest that the cockroach is much more likely than man to be the survivor in the future.

Since I have come to live at Hilltop, I have thought more about the species whose praises we do not sing. In fact, we do not usually take pains to protect most of them. Rather we do everything in our power to get rid of them. And many are doing very well. Moreover, they have succeeded in the face of one of the greatest threats animal life has ever endured, that from the killer biped, man.

I have noted the carcasses of some of these creatures on the side roads. I take trips on my bicycle every four or five days on the local roads. There are perhaps more dead animals on these side roads though plenty get killed on the freeways also. The most frequent animal I see dead hereabouts is the ground squirrel. Next are opossums, rabbits, skunks and coyotes. There are a few bodies of snakes also. Ground squirrels and snakes are day creatures, while the others are mainly night stalkers who get blinded by oncoming cars. It might seem that creatures which are killed by autos, particularly in the road culture of southern California, would hardly be from non-endangered species. I think though that the very fact that they are killed in such numbers is an indication of how successful they are. The countryside is so full of them that there is a constant spillover onto the roads. The number killed is insignificant compared to the total population. One does not see carcasses of the animals which are endangered because there are not very many. For instance, in five years I have seen only two carcasses of wildcats, while I have seen dozens of coyotes.

There is other evidence for the existence of large numbers of these animals in the countryside. I often see live rabbits and coyotes in the brushland between my house and the highway. While driving the one mile at night I have seen as many as five rabbits. Also, whenever I went down the hill to do some chore in the avocado grove, my two terriers had no problem jumping rabbits in the adjacent chaparral. And although because they are partial predators, they are not so numerous, I have seen dozens of coyotes on the hillside next to my house.

The night runners are like one another in several ways. They are widely successful throughout the country and in different environments. In particular, they do well both in the city and country. The ground squirrel, creature of daylight, does well only in the country and only in the arid west.

I had known of cottontails in the city and suburbs during my eastern days. They used to ravage my garden just outside New York City, and in Milwaukee they would gnaw off the bark of my young fruit trees in winter. I learned to put aluminum wrap around newly planted trees. It was no surprise therefore, when I found rabbits doing well in Los Angeles. Apart from the brush and parkland, several of the large airports of Greater Los Angeles had large populations of jacks while I lived there. I remember watching waves of gamboling long-ears at Los Angeles International and Long Beach Airports. Compared to the semi-arid brush land they came from, this was a paradise, a spread of level grassland to boggle a rabbit's imagination.

The coyote too has spread widely. Decades ago it crossed the continent from west to east, stopped only by the Atlantic Ocean. But it did not abandon its western homeland. Even while continuing to do well in the plains and brushland, it moved into agricultural environments, despite human efforts to get rid of it. After decades of trying to exterminate the brush wolf with all kinds of mechanical and chemical devices, the federal government has given up. The coyote has survived poison, the rancher with a shotgun, and the hunter in a Cessna, among other methods of killing. And being a true opportunist, it has found an even better place to live in the city. In both L.A. and San Diego coyotes course the ravines. They also thrive nicely in parks. What coyote from the Anza-Borrego Desert could imagine the delights of Griffiths Park, where besides the natural sources of food, there is much left behind by the two legged creature in whose midst the four-footed opportunist has come to live. The coyote also adjusted itself to the suburbs, which in Greater Los Angeles go on and on. There are considerable empty spaces between the settled areas. The coyote goes from one to another, using the brushland and ravines as home base and highway.

This survivor also dines on kinds of food it never found in the wild. It does very well on garbage, as is true of many of its brother survivors. One can drive at night along the ridge of Mulholland Drive between the city of Los Angeles and the San Fernando Valley and see coyotes foraging in garbage cans. Also the coyote quickly became adept at getting a new kind of prey, domestic cats and small dogs. These pets have not been sharpened for survival on their own as has their wild brother. Thus they become easy victims to the opportunistic hunter who chose to share man's space. Angelenos in areas where coyotes abound make a regular practice of getting their cats in when darkness falls. It is popularly believed that moonlit nights are particularly dangerous to small pets.

The threat exists in the country also. Here in North County very few cats are seen running free. Since I started this essay, my Himalayan cat pushed the screen open one night and got out. He was never seen again.

I have heard of several small dogs that were killed by coyotes, often very near the owners' homes. The dachshund, bred to face the fierce badger, falls easy victim to the nimble coyote. I worried about my Jack Russell Terriers when they were young, but it turned out they could take care of themselves against the pet killer.

Another animal that is adapted well to city life is the skunk. My first residence in California was in South Pasadena, only five miles from downtown Los Angeles. Except in parks, ravines, and the sloping sides of freeways, there are no open spaces left in South Pasadena. However, the skunk has found plenty of places in which to live. Under porches is a common location. At night in our backyard there was invariably the odor of skunk from animals going up and down the sidewalk. I never did see one. However, the skunk is almost totally nocturnal and is seen by us day creatures only rarely. I have never seen a live one in the countryside in North County either but I have seen plenty of carcasses on the roads.

Another medium-sized animal which I have never seen alive outdoors is the opossum. However, the carcasses of this creature are very common on local roads, particularly near creeks. I presume food is more available in such marshy areas. The animals stray onto the roads at night, are blinded by oncoming cars, and struck down.

The opossum too has learned to manage quite well in the city, and like other foragers, does well on garbage. Opossums were in the city already when I was a boy in Indianapolis. I also heard a lot about these marsupials in Wisconsin, where I lived for a number of years. The biologists at the Milwaukee Public Museum, where I worked, told me that basically the opossum was a tropical animal which had traveled far from its original home. In Wisconsin one of its problems was to get through the winter. It got cold enough there to frequently freeze off the tip of the opossum's naked tail. Therefore, I was little surprised when I found the opossum all over Greater Los Angeles and North County. Jessica saw one right in the middle of Westwood on her way to work one day.

The opossum is particularly interesting as a survivor. It is a marsupial like the kangaroo. Marsupials are generally considered to be mammals of a more primitive form. The newborn are in a very immature state and have to go through a long period of development in their mothers' pouches. Although once widely spread throughout the world, the greatest variety now live in Australia and the nearby islands. One of the bases of the theory of evolution by Alfred Wallace, the co-discoverer, was the line of separation between marsupial and placental mammals. It was called the Wallace Line. On one side placental mammals prevailed, on the other marsupials.

The placental mammals, like human beings, go through a much longer period of development within the mother and are thus born in a more advanced state of development. It is believed that placental mammals, like coyotes, skunks, squirrels and rabbits have replaced marsupials in most of the world. Australia and the nearby islands are unique because they were separated from Asia during the period when placental mammals were taking over. Kangaroos, koalas and wombats were not forced to compete with deer, racoons, and wolverines. The major exception is the opossum. It did not know it was a marsupial and second best. So, it spread over most of North and South America in successful competition with the placental mammals.

How do these survivors do so well when their relatives, the wolves, bears, wildcats, pandas, and monkeys do so poorly? I think there is one overall characteristic which helps the survivors --they can manage in a variety of conditions. The animals which are in difficulties are those which became too specialized. The South American monkey must live in the tropics. The opossum, while continuing to do alright in the tropics, has also moved into the temperate zone. In that sense its evolutionary history is parallel to that of humankind, which got started in the tropics but then moved to the temperate zone.

The wildcat does better in North County than other exclusive carnivores, though in far fewer numbers than the opossum. Where pressure from humans is not too great, the wildcat manages to survive. There is enough open country hereabouts for the wildcat to hunt small mammals and birds in relative peace. But it does not usually venture into the city, whether Los Angeles, San Diego or Escondido. The two wildcat bodies I have seen were adjacent to chaparral hillsides. I have never seen or heard of one in the city, although some of the very large, wild parks, such as Griffiths in Los Angeles, might have a few.

Once a creature becomes highly specialized for certain conditions, it is hard pressed if those conditions change. The South American monkey became highly specialized for living in the tops of trees. As those trees have been progressively cut, the chances for the monkey have become less and less. The baboons and their close relatives, the macaques, became the monkey generalists and have spread widely into different kinds of environments. The Japanese macaque makes out in snow country, other macaques make out in the mountains and deserts, while the majority keep doing well in tropical savannahs.

The ability to get along also depends on what an animal eats. Those which can eat many different foods are much better off than those which depend on one kind exclusively. The panda of China and the koala of Australia are endangered because they must have special kinds of leaves. Thus, except in zoos, they have to live where those plants grow. Not so the cottontail, which can do well munching the sparse shrubs of the chaparral in North County, the grass of the prairies, the bark of young trees or the grass of lawns In general, exclusive plant or meat eaters have a harder time finding food than all-eaters, or what have been more formally named, omnivores. Man himself is a clear example of the advantage of being an omnivore. Of the four most successful mammals in North County, all but one, the rabbit, are all-eaters. But the rabbit eats a wide variety of plant foods. They even come up to the house in the wee hours to eat my bird seed. The skunk, opossum and coyote will eat anything that is remotely edible, including garbage and new kinds of flesh. These creatures will not disappear because forests are cut down or a particular prey dies off or a hairless biped builds concrete jungles. They will merely shift to another food or habitat.

Then too, animals which are too easy to see are in trouble. They can be spotted too easily by the gun animal. Man is a day creature and generally looks for his prey between dawn and sundown. Even when he decides to hunt night creatures, he does so with artificial daylight, electric light. The famous four in North County are mainly night creatures. That means they are out mostly when man is not. Rabbits can be jumped in the daytime, but given their choice, rabbits will hide during daylight hours, coming out to forage when the sun gets low. The coyote basically uses the same part of the 24-hour day. The skunk and opossum are very rarely seen in the daytime.

Also animals that are larger are easier to see. The only good-sized predator which survives successfully in North County is the coyote. My guess is that this creature is near the maximum size to be able to easily disappear into the brush. Just about the only place I see them is crossing a road, and they disappear almost immediately when they enter the brush. The coyote's closest relative, the wolf, is long gone from this area. It survives only in remote corners of North America, having been eliminated by man elsewhere. And besides having more difficulty in hiding, the wolf is also more specialized in its diet than the coyote, being almost exclusively a meat eater.

The other big predators are not around any more either. There are a few bears in the mountains, but this animal is an all-eater, liking garbage and many plant products. Even so, bears survive mainly in national parks or forests. It is their size which gets them in trouble with the killer biped, not their diet.

The same is true of the grass eaters. The only good-sized one remaining in this area is the mule deer, which manages to survive with man's protection, though not in large numbers.

And although I picked basic survivors according to their body count on the local roads, I know there are many others. A number of birds are quite successful. The herring gull is an evolutionary winner. It eats insects, tidbits from the sea, other birds' eggs, fledglings and all kinds of garbage. It may fit Mormon notions of history to enshrine the gull as a gift of god because it came to eat the locust plague, but this feathered omnivore will probably go on long after the Mormons are gone.

Two other birds which have succeeded well are the house

sparrow and the pigeon. Both have spread throughout the world by traveling with the cultural biped. In the Midwest they do well living in barns, while dining on the spilled grain of their generally unwilling hosts. In the city they nest in the projections and crannies of man's constructions, while feeding on popcorn, bread crumbs, french fries and other human leavings. The small one came out of Africa as the weaver finch, and by attaching itself to human society, became the house sparrow. It poked its woven nests into the crannies of buildings instead of hanging them onto tree limbs, as its African ancestor had learned to do.

The Chinese had a campaign some years ago to rid the country of the pestiferous sparrow. Typical of the Chinese organizational approach, they set aside a day during which everyone was to make noise or wave harassing objects toward any birds seen. The idea was to make the sparrows drop in exhaustion, thus to be easily killed. An ingenious idea, though after their ordeal, the survivors got busy breeding again. When I was in China, 15-20 years after the eradication campaign, there were plenty of sparrows.

The pigeon's ancestor was the rock dove of Europe. It found masonry nooks in the cities good substitutes for crevasses on cliff sides. There was of course a lot more to eat in the towns also. There are not many pigeons or sparrows here in the country because there are not enough people and their droppings, nor grain fields. Pigeons and sparrows thrived at the university where I taught, however, clustering on the masonry projections and window sills, and descending to the quad to beg crumbs from students or pick up bits of fallen food.

However successful they may currently be, these two birds do have a built-in risk as survivalists. By being so closely dependent on man, they would be hard pressed if he were to be taken out of the picture. In the bird kingdom I would put my money on the crow. It does manage in the communities of men but also forages widely in the countryside.

And though most of us do not worry about their destiny, I think we must look at insects as potential inheritors. In fact, many are very well off. It is true that most have specialized diets, and most are adapted to particular living conditions. Most need special kinds of plant parts or prey, usually other insects. Also they all need a warm micro-climate. After what I said about the advantages of being a generalist, how can these creatures be regarded as potential inheritors?

What do insects have in their favor? First off, they are not highly visible. Then they have dozens and hundreds of young at one time. These grow to adulthood very quickly. This means they have more chances for basic biological change, mutations, than do creatures with long lives like elephants and humans. These latter also have only one baby at a time. Furthermore, it takes years for the infants to grow up and become reproducers themselves. Many insects reach maturity in days. As a consequence insects can easily become immune to poisons through genetic change. The ordinary flea, which has adapted to the bodies of mammals, including man's pets, does very well despite man's poisons. Jessica sprayed our dogs regularly in the hot season. The effects of the treatment lasted about one week before the flea population came back up.

When the insect can eat a wide variety of foods, it is a real biological winner. I have long been fascinated by the cockroach. To me, it seems to be a very likely inheritor of life on earth. Not only does it do well in the leafy layer of the tropical forest, but like the mouse and rat, house fly and sparrow, it has attached itself to man. Practically everywhere man has gone, the cockroach has accompanied him. It has substituted the nooks and crannies of the killer's residence for hiding places on the forest floor. Moreover, its ability to eat practically anything makes the cockroach a double threat. Besides droppings on forest floors, it has learned to eat leather, finger nails, eyebrows, book bindings and the glue on the inside of TV sets. It joins the opossum, coyote and rat in doing well on human garbage. It even has an ideal body shape, flattened, so it can squeeze into cracks when predators appear.

I had visible cockroaches in the last two houses I lived in, the first because I decided to share my place with the six-legged opportunists. I was convinced by my own lectures. In the beginning I did nothing to eliminate the insects except to keep the apartment clean. The creatures multiplied so rapidly that in a few weeks when the light was turned on, the floors and walls literally clicked with their scurryings. I resorted to poison then, man's primary method of eliminating other life forms. Here at Hilltop I got infested also, I suspect from a few survivors I carried along when I moved. Determining not to repeat the same mistake, I began an eradication campaign right away. It took a couple of years to get rid of them, despite cockroach bait left permanently in the backs of drawers. I am sure if I had stopped using the bait, that they would have rapidly built up their numbers again from the outdoor population, or from groceries.

It seems to me that in a day of potential nuclear destruction that the cockroach would be in a very favorable position. It has unlimited underground bomb shelters, the sewers of the world. If there were to be an all-out holocaust, the creature most likely to put its antenna forth when the ground cooled would be the cockroach. Of course, the rat and mouse would be close behind, and it wouldn't be long before the scavenging crow would come winging across the horizon.

There are still the vast numbers of microscopic creatures who are in no way endangered. These are the virus, bacteria, and other simple but reproductively very successful beings. Some are of enormous benefit to man, others attack every portion of his body. Much of man's vaunted technology is used to control those deemed harmful. Most are kept in check. But no scientist imagines that these beings are endangered. And when man's day is done, there is little doubt that most of these creatures will go on.

Is it possible to draw a profile of the kind of creature which will probably inherit the future, based on our current ideas of evolution? Of course, as humans, we wish it would be our own species. But when we look at the previous record realistically, we have to realize that anything so complex and specialized as man is probably not destined for long continuance in geological terms. Some simpler creatures will probably be the inheritors. I would guess that it would be those which could eat a wide variety of foods and live in a wide variety of ecological zones. Moreover, because of man's current dominance, it would have to be creatures with low visibility. And finally it would need to be a type able to produce large numbers of young which would mature rapidly. If I were asked to point out some candidates, I would start with the microscopic creatures, then the cockroach for the insect category, the opossum for marsupials, the mouse or rat for placental mammals, and the crow or herring gull for birds. The largest potential survivor I would select would be the coyote.

We should not be too surprised that the inheritors of life on earth will be inconspicuous creatures. The history of past eras tells us that is what happened before. The amphibian, the water creature that came on land, like the modern mud skipper, was truly insignificant compared to the large fishes existing in the oceans of the time. The small mammals could have been no more than nuisances to the great dinosaurs. And who could have believed that the lowly tree shrew could have been the forefather of the creature which learned to walk erect and use its forearms for tools? And of what significance were the talking bipeds to the great cats with whom they shared the African savannah? Many specialists think the first man creatures were primarily eaters of carrion. Being unable to kill large animals themselves, they fed off the kills of lions and leopards after the big cats had had their fill. It was the garbage of the era. Thus the early hominids were in good company with the coyote, opossum, rat and cockroach. But now after four million years of technological development, the leopard and lion will survive only if the tool-using primate permits them to. The insignificant two-legged creature has become the master of all large mammals. However, such is not the case with the small creatures. But it should be remembered that it was the inconspicuous ones which emerged in the past as the inheritors, not the spectacular ones.

So are we supposed to quit struggling against the natural forces? Hardly. I do not think that is likely, nor ever has been. There is little evidence that animals which were being replaced by better adapted ones willingly accepted their fate. The mammoth, elephant, bison, wolf and tiger all have tried to survive in the face of the man threat. So it is even less likely that man, with his hard-won technological knowledge, will accept replacement by some insignificant, possibly despised species. He will struggle with what weapons he has, primarily his technology, as long as he can.

But even so, it is well for man to remember that he is an integral part of the whole web of life. And besides the California condor, the grey whale and white rhino, this includes the coyote, opossum, house sparrow, cockroach and blue-green algae. Let man try to maintain himself as long as he can, but let him not be resentful at the successes of the insignificant ones, those who wait in the wings. They are relatives also, however distant. And finally let him not forget that some of them are the probable inheritors of life.

12

The Glorious Sun

    I live in one of the sunnier parts of North America, southern California. Also, I did all my field work in sunny, tropical places. I have watched the plants and animals which live around me adjust to the sun's power. And as anthropologists well know, man too has had to adjust to sun power throughout history. The second vital necessity for plant and animal life is moisture which also comes from the power of the sun. Even the tempo of life, from summer to winter, and from day to night, is controlled by the movement of the sun. No wonder so many cultural systems in the world had systems of sun worship. And if there were a viable religion of sun worship still on earth today, I could happily be a member, far more than that of a bearded homocentric deity, concerned only with the destiny of man.

    I can live with some cloudy weather. But if it remains overcast many days, I become depressed. I think my attraction to the sun, more than any other force, is what kept me moving ever southward during my lifetime. I still remember the long cloudy periods in the Midwest, usually with rain or snow. From day to day one could not count on greeting the sun.

    The sun does not always come up first thing in the morning at Hilltop, but most of the time it comes out by midday. There are few days when the sun does not shine at all. We get an average of at least

5.2 hours of sunshine a day throughout the year. The only places in the continental U.S. which get slightly more are the western deserts and the southern tip of Florida. I realize that there are some other factors which caused me to end up in southern California, but its sunniness is near the top. But though I know it has many other attractions, because of the dearth of sunlight, I cannot imagine ending up in the rainy northeast with its average of 3.5 hours of sunlight, much less the super-rainy northwest.

Even before coming to southern California, I had been moving toward the sun. When it came time for me to go into the field as an anthropologist, there was no question that it would be anything but the tropics. Other anthropologists went to the forests of the north to be with the Cree or Ojibway, while still others went all the way to the Arctic shore to be with the Eskimo. Not me. I had to be where frosts did not come, and where the sun was shining most of the year.

The part of the tropics where I spent most field time was monsoon Asia. Moreover, it rains quite a lot there. And I must admit that there were times in the middle of the rainy season in India and Laos when I became depressed. The monsoon clouds would sometimes hang low for days. But most of the year there was steady sunshine. I also spent time in tropical places where there was some rain most of the year. But climates such as those on the Caribbean Islands and West Africa tend to have frequent showers with much sunshine in between. They rarely have the long grey palls of northern low pressure systems.

There are people who cannot stomach the desert heat and aridity. I can understand this, especially if they have come from the verdant east. It can become very hot in the desert. My son and I crossed the country by bicycle some years ago, and I barely made it across the California desert. It was the most difficult piece of terrain we crossed, far worse than the Rockies. We left Twenty-Nine Palms as early as we could, expecting to reach Rice before the afternoon heat became intense. Due to head winds and heat, we did not. And because of exhaustion I could not make the last five miles on my own. Fortunately, a car stopped and took me the final distance. We later decided we should have done that stretch at night.

So the desert is the supreme sun region, getting even more than we at Hilltop. It is the only part of the country where one can effectively grow the queen of the oasis, the date palm.

With the elaborate technical devices of the twentieth century, the desert can be made comfortable. Palm Springs and Las Vegas are the best examples, though there are many other less famous desert communities. The people who live in those places are sun worshipers of a sort also. They simply go into their air conditioned houses or swimming pools when the heat gets too intense.

I used to have a fantasy of how I would like to end my days. It had never appealed to me to imagine dying while lying in a bed in a closed, darkened room. I wanted my last experience to contain much light. So I imagined myself walking across the rocky soil of a desert, carrying a bottle of whiskey. The sun would be beating down on me mercilessly. There would be no clouds. I would take a sip of my whiskey periodically to keep up my resolve, and while looking straight at the sun, yell at it in a loud voice. I couldn't make out what the words were but I knew I was trying to communicate with that great giver of life, using my last energies to get me into the sun land of eternity. Finally, I would collapse and the sun would dry me to a crisp. That seemed to me so much better than passage through pearly gates in a cloud land.

My former wife, who grew up in the verdant East and who was frightened by the desert, interpreted this fantasy as a death wish. When I was very ill and depressed once during a serious illness, she told the doctor that I wanted to die by being burned up by the sun. With no further information, he took this seriously and steered me to a psychiatrist. The fantasy was no more than intense rapture with the sun, certainly no wish to die any sooner than necessary. I got out of the hospital, dropped the shrink after three visits, and recovered from my illness on my own. I know it is unlikely that I will end my life in that fashion. It is merely the ultimate fantasy of a dedicated sun worshiper.

How do earth creatures manage with this powerful force? The sun existed long before the simplest form of life. Thus, living things have had to protect themselves from the sun, even while getting their sustenance from it. While the sun provides the basic energy for all life, it does so on its own terms. Plants and animals, thus, must have special equipment to take advantage of the sun's rays without being damaged by their intensity. Leafy plants take their strength from the sun's rays. In darkness, they sprout but do not grow. Instead they twist and turn on ever longer white tendrils, seeking light. If they do not get to sunlight, they collapse in sickly, twisted whiteness. Those which do grow have their own special needs for light. Gardeners know that shade-loving plants cannot be moved into full sunlight. And those which thrive in full sunlight will not produce in partial shade. I recently got some papaya seedlings from a nursery where they had been grown in a hothouse with reduced light. The nurseryman told me that I would have to harden them gradually to full sunlight. I put them in a shady corner outside. And though they got only 2-3 hours of weak afternoon sunlight, one of the four burned up.

Different animals are adjusted to particular degrees of light also. Since they can move about more easily, much of their adaptation consists of going to the proper place. Reptiles need warmth to be active. Thus at Hilltop, lizards get numerous only as the days lengthen and the sun's rays get stronger. Of course this also makes for an increase in insects, which means more food for the little reptiles. By midsummer, lizards can be seen every few feet, each scurrying to grab an insect, then hurrying back into the shade to eat it. Also the increase in insects brings the insect-eating birds. I watch the swallow nest under the garage eave closely as spring gets warmer. They have come each year as soon as there are enough flying insects around, about one week after they arrive at Capistrano. I think of them as needing an extra week to find their way into the country.

There are a fair number of snakes on the hill, mostly gopher snakes, though some rattlers and a few other kinds. I hardly see any from late November to April, our cool months. They are then underground. There are exceptional warm periods some winters. One January I was surprised one warm evening when upon returning afoot from the mailbox, I heard a slight buzz. That is the best description of the sound I can give. It certainly was not a rattle. I looked down and saw a baby rattlesnake coiled on the road near my foot. It was only about twelve inches long. There was no doubt, however, of its species. The head was the typical triangle, the body patterned like that of mature snakes, and the tip of its tail had a set of tiny rattles. They were vibrating, creating the buzz I was hearing. I gently pushed the little snake off the road with my shoe tip. It was lethargic, and though trying to twist into a striking coil, did not straighten. My guess was that it had just recently come out of the ground because of the warmth. But still it was January and the little serpent had not shaken off the winter chill completely.

Generally, however, snakes, including rattlers, do not appear in numbers until April or May, after the sun has risen more than halfway toward zenith in its sky crossing. Early in the spring, the snakes come out on the roads to soak up the sun's heat. In full summer they will not do that, rather seeking the shade during the day. Though they need the warmth of the sun, only so much will do.

Homo sapiens had to adjust to the sun's intensity also. A basic way has been through genetic change. In general, people in the tropics have darker skins than people of the north, a part of the primate heritage. A special pigment, melanin, gives protection. Another way of protection is the temporary darkening we call tanning. Light skinned people can get painful burns from the direct sun rays. Current medical opinion holds that skin cancer can be caused by too much exposure. Other things also get sunburned. Avocado trees, the trunks of which are normally protected from direct sun rays by their leaves, get burned if the tops are cut off. Grower's whitewash newly cut tree trunks until a new canopy of leaves has grown. In the trimming process, called "stumping," avocado trees are cut off at about five feet, thus leaving them capable of regrowing a canopy.

The human skin tries to adjust to the sun by producing temporary pigment. Northerners of European extraction who do not burn so much that they peel, get darker. They can get quite dark. On a bicycle trip I made with my son, we passed through the Navajo Indian reservation in Arizona. We had been on the road then for about two weeks, bicycling 10-12 hours a day, most of the time wearing nothing but shorts. On the reservation we stopped periodically at trading posts for cold drinks. As would be expected, most of the customers in the stores were Indians. Navajos, though not the darkest of Indians, are a dark tan. Once when we came out of a store my son said, "You know you were darker than any Indian in that store."

The fact that light-skinned people will get darker if they are gradually exposed to the sun has become a matter of status in Western culture. Persons with tans have enough leisure to stay in sunlight for long periods, usually doing nothing at that time. Thus a large industry in cosmetics to help avoid burns has been developed.

This attitude toward the sun is for light skinned people only. In the tropics, a darkened skin usually indicates being in a lower caste. People who work in the fields get darker because making a livelihood requires spending a lot of time in the sun. With what leisure they have, such workers are not to be found sun bathing. In fact, in many oriental countries female field workers carefully cover themselves so no rays will reach their skin. Dark-skinned ladies do a lot worse in the marriage market.

The main way man protects himself from the sun's intensity is through his culture, the inherited wisdom of his group. Unlike the other animals, humans have come to rely on learning for solving most problems. Adjusting to the intensity of the sun has been one such. And probably clothing has been the most important way.

Most people of the tropics are so dark genetically that skin protection is not necessary. However, northerners who have moved into areas where the sun is of tropical intensity have had to evolve some kind of protection. The most noteworthy example are the people from the northland, Europe. Even North Africans and Middle Easterners, who not so terribly long ago moved in to the region, have had to adjust to this climatic requirement. Thus they developed long, loose garments which protected them from the intense rays, while also providing some insulation. Also their skins became a little darker than north Europeans.

At Hilltop we have a fairly strong sun which shines most days of the year. The majority of the people who live in the hills hereabouts are of European ancestry and have light skins. Thus, the middle aged and older do not go in the sun very much. And when they do, they usually wear protective clothing .

Being of German ancestry, I too have a fairly light skin. Fortunately, it tans easily. And since I get outside two to three hours every day, I keep enough of a tan to avoid burning. When the weather gets warm enough, I strip to shorts for doing outside chores. I tend to get fairly dark then. In fall and winter my skin gets lighter. The people who get sunburned and cancer are those who spend most of their time indoors and then get massive doses of sunlight on a light skin in a short period of time.

True agricultural workers of whatever ethnic background do

not strip down even when it is hot. In the U.S., shorts were originally for city dwellers only. The Indiana farmers of my boyhood had no shorts and did all their field work fully clothed throughout the year. This usually included overalls, arm length shirts and hats. It is the same outfit that modern country western performers wear for showing a "down country" look. My father never wore shorts, even though he moved to the city in his twenties. His basic customs had already been formed by then. The skin of the bodies of my uncles were almost milky white, no matter what season. The term, "redneck" comes from the fact that the back of the neck was the only place on the rural worker's body that was exposed to sunlight. It got scorched. I suspect that most of the Anglo farmers hereabouts still cover up when they work outside. Of course, there aren't many who do other than overseeing. The man who works all day in the fields hereabouts is a Latino. And his skin is generally dark enough that sunburn is hardly a problem. Even so, most field workers remained fully clothed. Domingo, the Guatemalan who works at Hilltop one day a week, will work wearing pants and a tank top, but I have never seen him in shorts.

One kind of White man hereabouts works outside most of the time without being fully clothed, the construction worker. These young men get well tanned, wearing shorts and short sleeved shirts or with no clothing above the waist. Many get quite dark.

But apart from being able to burn, the great orb is also the absolute giver of life, here at Hilltop and elsewhere. Once young plants get hardened, they get their energy from sunlight. Water, chemicals and minerals are the other components necessary for their growth. But even water is carried by the sun, brought to us at Hilltop through canals and pipe systems, though originally from rains north and east of here. But rain could not occur without the evaporative power of the sun, nor could water run downhill without the force of gravity, a force which comes from the earth's relationship to the sun. The sun alone will not sustain life. Plants have to get some moisture. The desert plants get by with very little. Here at Hilltop we have the chaparral, vegetation of a semi-arid climate. We get about ten inches of rain a year, almost all in the winter and early spring. By May in an average year the ground on the uncultivated hillsides usually looks dry. But there is deep water and the native plants tap it through the long dry season, getting browner and browner, but surviving until the first drops begin falling again in the autumn. Even when there is no rain until December, the plants in the chaparral manage. There is more moisture in the bottoms of the canyons and more shiny verdant growth. The live oak and spiky yucca do well there.

The Indians hereabouts had a way of life that depended on gathering wild plants and hunting game. Their main specialty was the acorn, the fruit of the live oak. They had worked out a number of clever methods for processing this abundant wild food. But still this method did not produce enough for supporting a large population, certainly not a city. Except around their missions, the Spaniards got by utilizing natural rainfall. The Anglos brought in water on a vast scale which, with the abundant sunlight, turned California valleys and hillsides into productive gardens and fields. One can now grow an enormous variety of plants hereabouts, from those which originally came from the full tropics, to those which came from the cool zones. In the front of the house at Hilltop there are bananas and papayas, in the back apples and peaches.

The growth of plants controls the animal population. The snails eat leafy plants and the roadrunner eats the snails while the ground squirrels eat avocados and hawks prey on ground squirrels.

The sun also provides direct energy by giving warmth. The difference of seasons is simply that the sun is higher in summer, so the days are longer and hotter. I can grow tomatoes, eggplant and zinnia in the summer, while in the winter I must be satisfied with slower growing cabbage, cauliflower and calendulas. Even in the warm season, if the sun is blotted out by clouds, it gets cool.

This energy can be tapped. Solar energy devices are big hereabouts. In contrast to sun-created energy from past eras, coal and oil, daily sunlight power is continuous. In years past I had three panels on my roof, one for hot water and two for heating the spa. I also heat my house in winter with solar energy in the form of firewood. The first year I came I had a wood burning stove installed. It replaced the electric wall heaters which were put in the house originally. The stove keeps the house toasty warm throughout the winter. Wood, of course, exists because of the sun.

The tempo of life at Hilltop is controlled almost entirely by the sun. The wild plants flourish after the first rains and wither as the hot sun of summer dries the ground. Plants which are given water are another story. They grow luxuriously during the hot, dry season. The banana plants grow twelve inches a day during the summer. The winter garden is productive, producing large turnips, beets, cauliflower and broccoli. But the plants grow more slowly and less abundantly than the tomatoes, eggplant, peppers and zucchini of summer.

The rise in seasonal temperature marks the breeding season. Baby rabbits are to be seen in April and May. By April the birds are nesting, the phoebe building its nest on top of one of the flood lamps which hangs from the eaves, the doves in the forks of nearby fruit trees, and the house finches in the top of the yucca tree on the other side of the drive.

There is one pair of house finches which decided it would raise its brood on the front porch. This created a problem for me, as well as for the rose breasted birds. For two years now the female has decided that the Thai spirit house which I kept on the front porch would make an ideal nesting site. The spirit house was a memento of my days in Southeast Asia, a miniature wood dwelling. Thai families kept these model houses in their yards to placate the household spirits. They put cooked rice and other edibles on the little porches. I brought this one back many years ago and began the practice of setting it in front of my California houses, ostensibly to placate my household spirits. Hilltop seemed most appropriate. I set the little house on a stand on the front porch and started the practice of keeping a fruit of the season on the tiny porch.

All this meant nothing to the finches. The miniature house, with a tiny room covered by a pagoda style roof, was simply a fine place to build a nest. The birds first built the nest in the little room and dutifully deposited the eggs. The male would often stand guard, hanging on the chain of the porch swing, while the female sat on the eggs. Then trouble began. The spirit house was next to the front door and each time someone would go in or out, the finch would dart away. It abandoned its eggs, going to the other end of the porch to build another nest in a pot of ferns. It raised its brood there.

The birds did not have a long memory. The next year they built their nest in the same spot and the female laid five eggs. She soon began the same darting procedure. I decided to step in this time by moving the house to the other end of the porch. The mother finch stayed on the job. But another problem arose. I had put the stand too close to the wood box and one night a rat climbed the pile and helped himself to all the eggs. Then to add insult to injury, it gnawed up a portion of the avocado which had been put on the porch as an offering to the household spirit. Nothing was left in the morning but a couple of empty eggshells. The finches abandoned the nest again.

As the sun moves to the south in fall and the rays get weaker, there is a slowdown in life of both animals and plants. Reptiles and insects get scarcer, the insect-eating birds leave for the full tropics, other birds sing much less, and plant growth slows down. The rains come again in midwinter during the shortest days. The growth cycle begins again.

Not being given to traditional religious worship, I am a pragmatic sun worshiper. I try to use the power of the great orb directly and naturally. How could a grower of plants be otherwise? I have tried to become as self-sufficient at Hilltop as possible, and almost all my efforts depend on sun power. As described before, one of the first things when I came was to have the number of windows on the sun side of the house increased. A double glass door went into each room, facing east. This was partially for the pleasure of gazing out at the rolling hills. But it was also for letting the warming sun into the house in winter. From November until May the blinds are wide open. Of course the sun has to be kept out the other five months, lest the house become a furnace. The blinds are then mostly closed. Also I had a deck built the length of the house on the sun side so I could sit outside in the warming rays. Then came the wood burning stove and the solar panels.

I even have a filtration system which operates mostly through sun power. The pond next to the house is full of koi, decorative carp. No sooner had I cleared the water of chlorine and muriatic acid, than the algae began to grow and the sediment accumulate. Algae thrive best in full sunlight. Most koi keepers cover their ponds with fabric sun screens, besides installing elaborate motor driven filtration systems. My pond was much larger than most, and screens would have been a major proposition. Moreover, any screen would have cluttered up the view of the distant hills, which I valued highly. By asking around, I found out about water hyacinths, floating water plants, originally from South America, now in tropical and subtropical places all over. They reputedly cleansed water by trapping the particles. Further, they literally shaded and starved algae out of existence. In fact, experiments have been going on to use hyacinths for city filtration systems. I introduced some into the pond and they took off. And when 4/5 of the pond's surface was covered, the water became crystal clear. I put in a floating barrier on both sides of the bridge to keep 1/5 of the surface clear of the plants. Also I had a medium sized, simple gravel filter put in on one side of the pond. Thus the hyacinth, cleansed water gets a little gravel filtration before coming back into the pond. The floating plants grow like crazy from May to October. The excess must be forked out every week into small gullies on both ends of the pond. And the fish, now quite large and colorful, sparkle in the clear water, cleaned ultimately by sun growth power.

When I became an anthropologist, I came across the sun god in many ancient and exotic religions. Now that seems the most logical of beliefs. If there is any great power which is the source of all life, it must be the sun. If a religion still existed in which sun worship was central, I could comfortably take part in its rituals. If I were to have lived in another time and place I could easily have accepted Tonatiuh, the heavenly lord of the Aztecs, the Sun God of the Inca, Shamash the sun ruler of heaven and earth in ancient Babylonia, Re of ancient Egypt, the source of life and light, or the sun god in any of the numerous tribal religions. In the meantime, I do what I can as a practical sun worshiper.

13

From Cow Slaughter to Ahimsa

Apart from our genetic base, each of us is the sum of a series of events of a lifetime. So to explain a personality characteristic, especially if it is outside the norm, one needs to reconstruct the events that caused it to come into being. I do not eat red meat or birds in a culture in which the norm is heavy meat eating. I reconstruct that this came to be because: animals gave me more satisfaction than people in my boyhood, I worked in a slaughterhouse as a young man, I went to India on my first anthropological field trip, I quit hunting in my middle years, I lived next-door to a Chinese slaughterhouse in Laos, I killed several cobras on a field trip to Nigeria, and finally I was repelled by the sight of dismembered animal heads in a magazine in Switzerland. Of all influences I think my learning of the Hindu-Jain-Buddhist concept of *ahimsa* was the most profound.

It is a series of events in which the latest is a consequence of many which came before. It is only in hindsight that the final outcome is inevitable. But it is difficult for me to imagine the last occurrence without all the previous contributing causes. Thus, in order for there to have been X (the outcome), there had to be A, B, C, D, E, etc. (the antecedent causes) . If B or C had not occurred, then D and E, and all until X, would not have taken place.

We like to think there are single causes for our behavior, but this is a simplification. Yes, a man does not like women because he had a strong-willed mother who corrected him harshly when he touched his genitals. However, he was also short in stature and did not look very manly. He was rebuffed by several girls early in his maturity. But also he had three sisters and no brothers. And then after getting married, he was dragged into a very unpleasant California-style divorce. So who knows which was the particular cause for his alienation from women. In truth, probably all played a part, even if we do tend to pick one out as primary. And so it goes.

    I eat no meat except fish, which is outside the norm for my culture, that of the Burger People. And thus I am frequently drawn into conversations to explain my odd behavior. I have decided that a whole series of events and conditions have led me to this state of being. If any one of these events had not occurred, I might still be eating red meat. There may have been other contributing causes. One which is talked about much these days is for better health. Others have frequently thought that was the cause with me. And although my kind of vegetarianism may be better for the health, I cannot honestly claim that is why I gave up meat.

    I began my life in the Midwest as a member of the working class, one generation from the rural generation, but close enough to have inherited the standard American farmer's attitude toward animals. In general animals were and are classified as to whether they could be used or not. One learned enough about them to effectively use their flesh, strength, or other products, and one tried to eliminate those that couldn't be used directly. This is not to say that those hill farmers of southern Indiana, my uncles, never got a thrill from the sight of a wild duck or red shouldered hawk. But from what I remember, their most usual response was to take a shot at it with a 12-gauge shotgun. And I followed suit. As soon as I was permitted, I picked up a gun and began shooting at some of these creatures. All my relatives and most of my friends in the city were gunmen. And I continued shooting at wild life for twenty years, as circumstances permitted.

    My family had become urban well before my birth, and so despite farm visits, I grew up a city boy. But I spent some of the happiest times in the city along the railroad sidings, in the brush by the city dumps, and around abandoned gravel pits. When I got my first bicycle, I ranged farther.

I also kept many animals as pets in my backyard while my father used other animals in the customary exploitative way. He raised chickens and rabbits for food. One of the most unpleasant jobs I had then was holding the legs of freshly killed rabbits while he gutted and skinned them. Many farm children have been put in a similar quandary, usually accepting the necessity to send their 4-H Club animals to the slaughterhouse, directly or indirectly.

At this same period of my life I picked up some of the values of working class America, primarily an interest in automobiles and girls, and the need for money with which to get them. And not having any special training, I went from job to job during the Depression, trying to find something that would pay more. Thus, I got the worst possible job, work as a member of a clean-up gang in a slaughterhouse. And apart from it being the dirtiest, and probably most unhealthful job I have done, it was the center for the killing, dismembering and processing of animals. And though there were a number of aspects of the job which bothered me, I did not abandon the customs of my enculturation. I continued to be a staunch eater of bacon, ham, hamburger and steak.

The years passed and I became entrapped in the net of anthropology. A totally new world opened for me. Before long I began field work. And like others in the field, I started by following romantic notions from my earlier years. India, the exotic, beckoned and I decided to become a specialist in the cultures of the Holy Mother Ganges, the sacred cow, and the god-hooding cobra. I think the mark from that field study was deep. It isn't that I changed overnight, but now the more I look back and consider what I became, the more am I convinced that India and the Hindu concept of *ahimsa* (sacredness of life) have become a part of me. This was especially true when in answer to Jessica's call to help from an invading "elephant," I would catch the spider and put it outdoors; or when there was some leftover sugar in the bowl I would sprinkle it out under the eucalyptus trees for the ants.

From the first time I went there, India was a fascinating country, and no small part of the attraction was the super abundance of life, human and nonhuman, wild and domestic. Only there did I have whole families of mongoose stretched out on my tile roof, warming in the winter sun, great flocks of green parrots and flying foxes in the shade trees outside my house, gypsy entertainers on the streets with dancing bears or monkeys and pet gazelles in the workers' neighborhoods. And in the countryside, the life forms were even more varied and abundant. All this, I soon learned, was because the normal Hindu did not blast the creatures with a shotgun.

Of course there were problems from all this abundance of life. One of my first culture shocks in India was from all the mice. They lived abundantly in the walls of the first quarters arranged for me. And although I had seen plenty of mice before, I had never been so close to so many. They kept popping out of the numerous holes in the wall, staring at me with their shiny, obsidian eyes, then popping back in. And then just below the second-story window of the apartment where I was located, there was a Hindu temple where cows and people were continually entering, the worshipers leaving offerings, praying and ringing a gong, and the cows eating the offerings, or just standing around, chewing their cud. That went on all night. I didn't last there many nights, but neither did I manage to stick it out at the government bungalow on the edge of a village which had been arranged by a well-meaning Indian bureaucrat I had met. How could he guess in advance that the sounds of jackals that night would unnerve me to the point of fleeing the next morning.

I survived the initial series of shocks and went on to do my field study -- in a city of one million people where only a Hindu mathematician could compute the number of animals. And I kept encountering creatures during my year and a half stay. Perhaps the most memorable events were the rat which woke me in the middle of one night by pulling my hair and the pack of rhesus monkeys which drove me into the latrine of a college hostel where I was staying. These encounters I also survived, and at the end of a year I was comfortable with the varying life forms, so characteristic of a Hindu environment, so alien to a Christian one.

Laos, the next place where I spent several years, was more typical of a peasant society, no matter that the locals were Therevada Buddhists. Not only were the Laotians not vegetarians, the typical villager killed and ate whatever he could lay his hands on. There was little wildlife within walking distance of the villages, though not too far beyond there were jungles which still harbored wild cattle, elephant, tiger and leopard.

The one wild creature that still managed not too far from the villages was the Southeast Asian jungle fowl, the ancestor of all domestic chickens. I must ashamedly admit that I hunted these creatures, though in my defense I never got one and gave up hunting just after that period of my life. My method of stalking them was very inefficient. I would creep up on the sound of a crowing cock. But invariably just before I would get the flock in sight, the crowing would stop and when I would take a few steps more, the forest floor would burst into a blur of whirring wings. I am of course now pleased that I never got a shot at one of them. Before I left the country, I gave away my twelve gauge, never to use another. The Lao villagers mostly used crossbows, snares or nets to catch local wild creatures. They must have raided nests also. I used to go to the local markets where I would invariably find some captured jungle creatures.

Perhaps one of the most interesting ones I bought was a slow loris. I remembered the small pre-primate from a class in physical anthropology. Mine was a wide-eyed creature, a little larger than a squirrel and which justly deserved its name. The fastest movement it ever made could justly be called a slow crawl. I fed it bananas, which it would squeeze into paste with its little humanoid hands, while staring at me unblinkingly with its enormous, night-seeing eyes. When it wasn't eating, it mostly slept, the night being its element. I am happy at the remembrance of releasing it back unharmed into the forest. It stared at me in its usual unblinking way and slowly crawled up the tree trunk. May this tiny relative, or its descendants, still roam the remaining trees.

As with most subsistence farmers, the Lao considered these small creatures as food. They sold them for a pittance and I suspect they normally put them into curries or stews. I quit going to that part of the market after my purchase of a pair of pygmy owls. These little birds, about the size of quail, were trussed together and lay on the ground tranquilly when I first saw them. The asking price was almost nothing, so I bought the pair and took them home. I attributed their quietness to shock and rough treatment so was not overly perturbed when they did not get up immediately. However, when they did not get up for many hours, I asked my Lao assistant, a village man, if he would look at them. He quickly showed me what was wrong. The tiny ankle bones had been snapped so the birds couldn't claw people. I remember thinking it was similar to what was reported of the Carib Indians. Being cannibals, they would break the leg bones of their human captives so they couldn't fight or run away. After all, since they had been caught for food, this was the best way to keep them alive but harmless. This idea also made sense to the village farmer who had caught the owls and who was selling them for food. But to me, a Western city man who was tarrying only briefly in this village land, it was sad to contemplate the little mutilated creatures.

I have thought many times about the seeming callousness of the village farmer and the nomadic hunter toward the animals in their territory. This has made me sad as a Westerner imbued with some liberal concepts. But I have learned to accept as an anthropologist the necessity not to judge others as I would my own people. It is easy enough for the Western city man to condemn the African or Asian poacher who is killing gorillas or elephants. But when one looks at the economics of that subsistence farmer, one quickly comes to the realization that the money obtained for a pair of ivory tusks or a gorilla's head might mean a great deal, a bicycle, or a transistor radio or even some improved seed. I am not approving the killing of these magnificent beasts, or any others on the endangered list, but I do think those concerned need also to consider the needs of the poacher. The poor fellow needs an alternative. Condemnation and police action are not enough.

Anyway that was not my problem in Laos, even though spectacular wild animals were being killed off there also. To me at the time the proximity of a Chinese slaughterhouse was much more important. As an employee of the U.S. government, I and my family had been provided a very comfortable French colonial house in the town of Pakse. We had enough space for geese and turkeys in our yard while across the street my son could go pony or buffalo riding with his Vietnamese buddy. One negative was the fact that the empty lot just across the other street was used as an open air lavatory. The locals, practically none of whom had their own toilets, would come to squat in the bushes throughout the day. When the wind was wrong, the odor was high. But the most significant negative for me was the slaughterhouse about a block down the street. Late afternoon each day the Chinese would start hauling in trussed pigs on bicycle rickshas. Then at about dusk the squealing of pigs being stuck would start, to continue until the wee hours. Many nights were sleepless for me, I listening to the life-draining squeals, hoping the carnage would end.

However, I did not quit eating pork or other meat then, as I had not when I had worked in the slaughterhouse in Indianapolis. In fact, during that earlier job I had even gone to watch pig sticking deliberately. I had also forced myself to witness animal sacrifice on field trips later, particularly in Muslim countries. I had been taught that anthropologists were supposed to observe whatever happened on the local scene. Not to would indicate cultural bias, which was a professional no-no. The Chinese slaughterhouse particularly bothered me, I suppose because it lasted so long each night. It didn't help much that I accepted as an anthropologist the Chinese custom of using pork throughout history in a very efficient manner.

Some years later, wanting to broaden my cultural horizons, I took a research job in Nigeria. The area of Biafra where I went was intensely cultivated and there was little wild life around. The Ibo villagers were like most peasant cultivators in respect to wild creatures. They ate them. In fact, they had cleaned most out long before I got there. Thus, my experience with animals would hardly have been memorable had it not been for the cobras I killed. I still feel guilty about that though I get some relief each time I catch or guide a rattler away from the road. Here at Hilltop they come out to warm up on the pavement on cold mornings. Most of my neighbors run over them with cars.

Our house in Enugu, Nigeria was in a well-tended university compound. Chain link fencing had been put around a larger area which was not tended by gardeners and had grown up in low bush. Because villagers were kept out by the fence and guards, the brushy area supported more wild creatures than other nearby places. The campus compound in the center was well tended by grounds keepers but that stopped at the fence, a stone's throw from our house.

As I have done in most places where I lived, I started a little garden next to the house. It was doing well with marigolds and zinnias. One afternoon I was sitting behind the sliding glass doors of the living room when a dark snake slithered across the open area and disappeared in the flower garden. And although it had come rapidly, I felt sure I had seen a widening behind the head. I went out the back door where I had a long bamboo pole. Bringing it around the house, I stuck the pole in between the flowers. The black, hooded head of a cobra raised itself above plant level. I thrilled at the sight, my first wild cobra. And although it was obviously a young one, all the media biases of a lifetime swept over me. I suppose I felt courageous as I faced the poor creature. It raised up to get a view of me, the fearsome predator. I came down on it with my pole, breaking its back. In retrospect, how can I take pride in destroying that beautiful creature which meant me no harm? It must have been hunting for small rodents or insects. I killed three others in the yard later and one in a drainage ditch away from the house. All were half grown, I suspect from the same birthing. I boasted to my colleagues in a classic enactment of my role as a prime member of the great killer species, *Homo sapiens*.

Now, these many years later I regret this last killing more than all those more bloody ones of my callous youth. And though there are other aspects of the Christian heritage that I do not now find convivial, I particularly dislike the snake hatred that drove me to kill those creatures. In this regard I am glad that I do not believe in an afterlife, and particularly one which might have been designed by Hindus. For I can imagine that when I would go there, I would be met by the ghosts of those cobras, who after justifiably accusing and condemning me, would proceed to administer some particularly horrible reptilian torture. Perhaps I would be forced to stand weaving from side to side throughout eternity while a cobra guard would control my movements by slight nods of the head. After all, Hinduism is the only major religion which puts the cobra on a pedestal.

As the years passed, I became less and less willing to kill animals or to enjoy eating parts of them. And there came a day when I decided not to eat animal flesh any more. I had by then reached middle age and had achieved a minor reputation as a specialist in cultural change. I had got appointed as a consultant to a United Nations committee on protein deficiencies in the third world. In this capacity I went to a conference in Geneva which was less than memorable. However, when the meetings were over, I took a few days holiday. I went to see Chamonix, the Swiss mountain resort, a train ride away. It was cold and snowy, most appropriate for viewing the mountain. My return to Geneva required a wait of an hour or so.

After warming up in the train station, I took a walk down the platform. There in the cold, blowing wind I chanced across a magazine kiosk. In studying the display, I came to a picture journal whose cover had a cow's head, decorated with a flowery, spring hat. It repelled me in a fascinating way, the irony of decorating humanly the detached head of an animal killed for food. I picked up the journal and leafed through the pages. Each had a different animal's head with a different hat, a horse with a homburg, a goat with a sombrero, a sheep with a baby's bonnet, etc. What were presented as humorous titles were in captions below. To this day, I cannot say why this sight was so repulsive to me. I knew that the Swiss were as adamantly carnivorous as Americans. Also, I knew that throughout most of human prehistory man was a hunter. Furthermore, once the animals were dead, what did it matter what was done to their heads? But nothing helped. I continued to be totally repelled.

And as I walked away from that kiosk, I resolved never to eat animal flesh again. Actually this change of habit was one of the easiest I have made. It took me at least ten years to give up smoking and I still work at cutting down my consumption of alcohol. Switching to vegetal food has been easy by comparison. Furthermore, there have been some very good consequences. I think I have felt better, both because I knew I was not causing the death of other creatures and because my nutrition probably improved. With all those veggies, one cannot help getting more fiber, vitamins and minerals, as well as cutting down on one's cholesterol. And in our culture of over consumption, who needs the amount of protein the normal meat eater gets?

Also in my later years, I married a lady, Jessica, who was a vegetarian, and for most of the same reasons. She made it much more pleasant to get along in this culture of carnivores. When we met, neither she nor I had been searching for a vegetarian mate, but this shared characteristic did not hurt.

So, the Hindu idea of the sacredness of life has influenced me to become mostly a vegetarian, to be more and more reluctant to take animal life. I still eat fish and eggs. I gave up hunting and have not cared for guns for a long time now. But other behaviors seem to me to have been a consequence of that influence too. For though I have no missionary zeal to convert others in their treatment of living things, I do get bothered with heedless killing, largely a product of the Judeo-Christian heritage. After all, the creationism of this belief system effectively disenfranchised non-humans by denying them a soul. They were put on earth for man's benefit, and if he was not able to use them, he was free to destroy them. They had no inalienable rights. Thus a lot of mindless killing goes on. It makes me unhappy to see insects, snakes and small mammals killed around Hilltop, especially when they are doing nothing except being visible. How different from an orthodox Jain, a people who refused to continue farming because they might kill ground dwelling creatures unwittingly. As a consequence Jains took up occupations where there was little chance of killing anything directly. So by a twist of fate they became money handlers and some of the richest men in India.

I am not suggesting that any culture can continue without someone taking life, either animal or vegetal. It is obvious that someone has to farm and most Indians, including the Hindus, do. They produce the food for the much fewer Jains. But the Jain example indicates how far in the other direction people can go in respect to taking life.

On occasion, and to Jessica's amusement, I followed one of the reported practices of the Jains at Hilltop. Jain women would feed the ants in their gardens, sprinkling grains of sugar on the ground to which the accustomed insects would eagerly scramble. While not making a daily practice of this, I have poured the residue of sugar containers on the hill behind the house rather than dumping it in the disposall. And while Jessica hardly got her practices from Hindu-Jain beliefs, she did learn also not to kill wantonly. The one exception was insects, and particularly spiders. And although it is not unusual for women in Euro-American culture to have negative feelings about insects, hers was even stronger. She had substituted the word "elephant" for spider, never using the latter. Something in her past history had created this near phobia, as my attitude toward animals has been influenced by contact with Hindus. In any event, I early became accustomed to cries of alarm when she began coming to Hilltop. She would inform me from a distance that there was an "elephant" in such or such corner or on a place on the wall and would I eliminate it? She at first suggested vacuuming them up, but I quickly suggested an alternate solution. I would catch the creature, either picking it up by one of its legs or by using a paper napkin. Then I would carry it to the nearest door and put it outside. I don't know when exactly I gave up the usual practice of squashing insects but probably here at Hilltop. In any event, I feel much better for it. And Jessica never wanted to kill the dreaded creatures either. She merely wanted them gone.

It might be asked how I could be so sure that my anthropological background, and particularly my experiences in India, could have been so instrumental in forming my personality. In particular, how could that have turned me vegetarian? The answer is that I don't know for sure, but it seems to be the closest to a prime cause.

A personality is an interwoven tapestry made from one's genetic potential and life experiences. But since I don't know what my genetic potential is, I must concentrate on experiences. And I can see that the other happenings have been influential, particularly those of my boyhood. But I must face the fact that I continued to follow the practices of my Midwestern carnivorous ancestry until after the accumulation of several anthropological field trips, the most memorable being among the Hindus. And though I appreciate much else that I got from them, nothing is more important to me now than vegetarianism.

14

Is Peanut Butter Food?

There are many brown men from south of the border who come by Hilltop. The ancestors of these Latinos once possessed this land, having dispossessed other brown men, the Indians. Now we pale-skinned men hold the chaparral hills and valleys, and we have planted avocado and citrus groves on the sides of some hills. It is beautiful country, made the more so by the green groves and scattered fields of flowers and vegetables. The brown men must feel comfortable in this landscape, which is similar in many ways to the land they came from. Even the fruit which does so well on the hillsides, the avocado, is completely familiar to them, having been domesticated by their ancestors a few thousand years ago.

Small groups of these men, mostly in their twenties, come up the road every week or so, looking for work. They wear used clothing from yard sales or flea markets. Many top off the costume with a baseball or porkpie cap. Sometimes one of them carries a stick to ward off the dogs.

They usually stand in front of the door and say one of their dozen English words. "Work sir?" they ask.

Very few have the temerity to knock on the door. Some haven't even learned that single most critical term, and will ask in Spanish, "Trabajo?"

I usually answer them, "Ustedes hablan ingles?" (Do you speak English?)

Most often the leader will answer in his own language, "Un poquito, senor." (A little, sir.)

I then, "Yo hablo un poquito de español tambien." (I speak a little Spanish also.)

There would then be a noticeable relaxation, after which we would continue in Spanish.

Most of these men have come from Mexico or Guatemala and they are the lucky ones. They have already passed through some ordeals, including getting out of their villages alive if they came from Guatemala. Death squads have been reported to exist there which hunted down villagers. This is of course hardly anything new. Indian villagers have been savaged ever since the days of the conquistadors, when all Middle American nations were destroyed by the Spaniards. Mexicans who have got out recently were usually escaping nothing more serious than grinding poverty, a heritage of the Spanish conquest also. These villagers had to pass through some part of Guatemala and/or Mexico and finally make the tricky border crossing. Border crossing specialists, called "coyotes," guided them across the Mexico-U.S. border for $100-$150. Many had to try several times. I talked to one pair who had tried four times.

Once across, the immigrants had to decide where to go. Many went to the big cities, San Diego and Los Angeles, and beyond. One sees them every day, standing at special intersections, waiting to be picked up for day jobs. But many fan out into the countryside, to become the backbone of agricultural labor in California. We get our share in northern San Diego County, where they pick avocados, strawberries and citrus fruit, the bending, stooping, lifting jobs of migrant labor.

They live where they can find places, some in makeshift structures called spider holes, others in field hands' dormitories, while a lucky few have trailers, and even fewer have small houses. They buy from our local markets, choosing from the cornucopia of the U.S., the abundance that makes obesity a national problem.

Thus, one might imagine that they would eat anything thought by anyone as food. Not so, they also have their customs, and thus their preferences, their cultural baggage.

When they come to my house, I put some of them to work if I have any jobs that do not require too much supervision. They work hard until noon when they come to the house to be fed. I assume this is a carry-over from the worker-patron system they have in their own countries. Anyway, I early accommodated to this system and provided them a "lonche."

I spent some years alone on the farm, during which time I would fix this meal. I would put on a tray bread, cheese, jalapeño peppers, avocados, mayonnaise, jam, and a bottle of beer each. The items would be in their original containers, along with knives and spoons to spread or dish them out. It was a lazy, bachelor's way of doing things, though it seemed to work out well enough. The men picked out what they wanted, which was most of the items.

Then I got married and brought Jessica down to the farm for days at a time. Though a dedicated city girl, she enjoyed most of the experiences on the farm. On one of her first visits, she offered to take over the "lonche" chore. Jessica was a very caring person and also in some ways old fashioned. She always enjoyed fixing food for others and early on had taken over this job. It seemed natural that she would want to do so for field hands.

She fixed up a tray in fine style with cutlery, napkins and the works, similar to the ones she fixed for our meals on the deck. However, since she and I are vegetarians, there was no meat available. I did have a couple of cans of tuna, left from the days when I was still eating fish. At her query, I said it would probably be familiar enough. So it appeared on their separate plates as classic Anglo-American tuna fish sandwiches, made up of mayonnaise, pickles, onions, lettuce and tuna. It would have been fine for a North American style picnic. I thought the whole ensemble looked so American that I took a picture of it, the brown face of the Latinos watching in puzzlement. Still they ate their sandwiches with gusto, along with the side items, washing it all down with beer.

A couple of weeks later another group came and there was no tuna left. Jessica suggested feeding them peanut butter. I responded that peanut butter was a very American item, with which most other peoples were not familiar. I remembered seeking it vainly in France at my step-son's request some years before. I knew that the problem was not that other people did not eat peanuts. Malaysians and West

Africans delighted in peanut sauce which was hot and spicy and in which meat was usually simmered. Moreover, the peanut had been domesticated by the ancestors of these very Latinos in Central America. However, most Americans ordinarily ate peanuts as a snack food or in the aforesaid sandwiches. To them, goobers in a hot sauce was not appropriate. Moreover, spicy food in general was not preferred by North Americans.

Anyway, when Jessica started with an explanation about how nutritious peanut butter was, I told her to go ahead, to fix whatever she wanted for the Latinos. I had already noticed that there was cheese, jalapeños, beer and some other items on the tray. She fixed the sandwiches and I went off on some errand. When the workers were finished, they went back down the hill to finish their weeding job.

A short time later, I noticed our Dobie snuffing and pushing a shapeless brown conglomerate near the edge of the hill. I thought it was some small carcass, the bones having been mushed up by the dogs. The Dobie and rat terriers were always bringing in the remains of some creature they had caught.

I went to investigate and was surprised to notice purple globules of grape jelly on the light brown spread of peanut butter on top of a piece of bread. How much had been worried off by the dog and how much eaten or pulled off by the Latinos I will never know. One thing was clear, most of that peanut butter and jelly sandwich had been discarded. Perhaps the other one had also.

I took away what remained, since the dog did not seem interested anyway. Holding it behind my back, I carried the remains to where Jessica was sweeping the deck. "You'll never guess what I've got."

She studied me suspiciously in the way that girls who had brothers who put insects down their backs will. I assured her that it was harmless, which only partially alleviated her suspicion. I know that as a little girl she had been teased so long that she could not easily come to terms with any male who spoke softly but kept something hidden.

I held it out. "Look, know what it is?"

She stared for a bit in puzzlement, then with beginning awareness. I said, "It's the peanut butter sandwich. They threw it away."

Like a good dedicated ethnographer, I couldn't help asking the Latinos for their explanation. For as I have already explained, all peoples had explanations for everything they did. I interrogated them patiently and finally one answered shyly, "It looks like baby shit, señor." And so it does, but only to a Latino who is not accustomed to eating it, not to the North American who learned to eat it by opening a plastic baggie in his school lunch, there to find the familiar peanut butter and jelly sandwich.

At first I thought the affair was humorous. Perhaps it evoked memories of my reaction recently in Malaysia when an American group I was with was desperately trying to get some familiar foods, lemonade, cheeseburgers, cold milk and cheesecake from waiters who knew only a smattering of English and practically nothing of American customs. The Americans were particularly disillusioned because the hotel restaurant was featuring "American cuisine." And though they were serving steaks, roast beef and french fries, none came out right. The cooks were Malaysians or Chinese with little first-hand knowledge of how American dishes were prepared.

When I think back, I wonder what else could have been expected. Most people try hard to get familiar foods, as they try to follow other familiar customs. I had long accepted this process as logical. It was what kept a culture going. A culture in which everyone was trying to do something new would disintegrate rapidly. Unfortunately, it has always been difficult for me to accept the traditionalism of my own culture emotionally, even if I could do so logically.

Now I do not think the incident with the Latinos was funny, not even the image of the Dobie pushing the discarded piece of bread around. Everyone, man and beast, was acting according to his own upbringing. And after all, Jessica in her caring way was trying to help. I knew that unless seriously provoked, she couldn't be unkind to anyone. Even when she also was trying to get familiar foods in Malaysia, she never became angry with the waiters, as some of the other Americans did. That she had tried to feed the Latinos peanut butter and jelly really only means she doesn't always recognize that cultural differences exist. And in this, she is following the majority of humanity.

I should also have remembered the little exercise I had gone through in so many introductory anthropology classes to show that most of the people of a culture think and act the same on basic issues. When I would ask the group what were the usual foods sent along in school lunches, the invariable reply by at least eight out of ten was "peanut butter and jelly or baloney sandwiches." At least two generations of Americans have grown accustomed to these two items as standard school lunch fare. And this is not customary in all cultures, not even those of the West. My stepson switched to hard-crusted French bread, butter and jam when he went to a primary school in France. Bologna and peanut butter were not available.

And so to answer the question, "Is peanut butter food?' it must be to North Americans. But just as logically it is not food to Latinos and most others of the world, since they do not eat it. So perhaps it makes sense to change the question to "What is food?"

And on the basis of peanut butter, boiled dog, roasted grasshoppers, hamburger and milk, all of which are relished in certain cultures and shunned in others, I will offer a definition: "Foods are those substances which provide some nutrients and which are approved as edible in some cultures."

Of course this includes almost all the potentially edible substances on earth, but each only in some cultures. Thus there are no universal foods, only ones specific to particular cultures.

We anthropologists are one of the only breeds which keeps insisting that customs are relevant only to the culture where they are practiced. We do not accept the idea that there is any universal right and wrong, justice or injustice, method of greeting, or right kind of food. We keep insisting that culture is real.

So perhaps Jessica is right when she says that if I go first, this is what she will have inscribed on my tombstone:

CULTURE IS REAL

15

Love Affair With Avocados

I long ago became deeply involved with that marvelous fruit which flourishes so well in southern California, the avocado. From the time I first tasted the buttery flesh of this gift of the ancient Mexicans, until I completed the love affair in my avocado grove in North County, this pendulous green fruit and my existence were closely inter-related. And although I am very fond of many kinds of fruits and vegetables, the avocado is one apart.

Besides the enjoyment of eating it, my personal life became deeply involved in my relationship with the avocado. And although I am not a person inordinately given to ritualism, the avocado ultimately became a ritualistic symbol to me -- of life, love, growth? The ritualism was inspired by the first great love of my life, anthropology. Moreover, the two, the avocado and anthropology, became closely intertwined through the years. And in retrospect it seems almost inevitable that they did.

How did it all come about? As best I can remember, the following is what happened from the time I was introduced to the "testicle fruit" (the meaning of the original word for the fruit in the language of the Aztecs) until I performed the fertility ritual in my avocado grove north of Escondido. On my second field trip as an anthropologist, I was introduced to the fruit on the island of Trinidad. I first enjoyed the taste of many kinds of tropical fruits there, but the avocado (called the alligator pear in that region) was king. I couldn't get enough of it. After that trip, I traveled to South and Southeast Asia, as well as to Africa, but in none of those places did I find avocados. I made do with mangos, papayas, bananas, and other tropical fruits. Then my academic career brought me to southern California, first to South Pasadena. Having been a hobbyist gardener for years, both in the northern zone of the U.S. where I had had various academic positions, and in the tropical countries where my field trips had taken me, I quickly became hooked on the abundance of plants in southern California. I just couldn't believe the variety of fruits, vegetables and flowers which flourished year round. I quickly became aware of the fact that if you applied water to plants in southern California, you had better step back. Another early realization was that the avocado flourished in an abundance I had not seen since Trinidad. There were trees all over Los Angeles and its suburbs, some in yards, some on the grassy strips between the sidewalks and streets, and some abandoned on the sides of freeways.

Our first year my family and I lived in a small rented house in South Pasadena. I quickly learned that next door was a mature, producing avocado tree. I didn't know the varieties then, but I now know it was a Fuerte, probably the second best of the commercial varieties. I watched that tree hungrily for some time until when returning home one day, I found a good sized fruit on the sidewalk, where it had obviously rolled after falling from the tree. (I also learned later that such fruit were known in California as "rollers.") I picked it up and noticed small tooth marks around the stem. Later I learned that stems were frequently gnawed off by rats, allowing the fruits to drop. I took it home anyway and trimmed it around the gnawed place, eating the fruit later when it had ripened. That started me off. I got into the practice of going out early each morning to pick up fruit that had fallen into my yard (the tree next door hung over it). I felt a little guilty since I had not asked the neighbor if he minded. Later I learned that he didn't care at all. Like many urbanites in southern California, he was so used to avocados and other fruit, that he did not bother to pick all that his tree produced.

I developed another habit in those days to harvest some of the fallen fruit of my neighborhood. I was then riding my bicycle each day to the college where I was teaching, and I learned that quite a few overhanging trees grew along the back streets. I modified my route accordingly. As on the tree of my neighbor, rats would gnaw fruit loose during the night. I would pick them up in the morning as I passed, usually getting two or three fruit each day.

At about this time an occurrence was taking place in my personal life which would contribute greatly to my growing relationship with the avocado, my marriage was breaking up. Of course I now realize that marital break-ups are as Californian as avocados, so a linkage seems most natural. The Golden State is now the land of brittle marriages, as well as being one of the two subtropic garden spots of the U.S. And so, like so many other middle-aged Californians, I went through my first divorce. And as with so many others, I became a member of that vast amorphous body of adults we came to refer to as singles. Moreover, despite my advanced years, I began dating again. But no matter what activities I engaged in, I did not neglect my interest in avocados. Whenever I had a chance, I engaged in my new found sport, harvesting avocados. I usually could involve my lady friends in the search also. At the very least, they all liked to eat the fruit.

I remember one late evening when a friend of mine by the name of Martine and I were invited to have dinner with a graduate student-friend, who I had learned earlier had a couple dozen producing avocado trees in his large yard. We were to have dinner with him. His wife was off visiting relatives. I had got to know Ed originally because he was a student in our department. However, I am sure his popularity had gone up considerably when I had learned of his avocado trees. Whenever I visited him, I could count on taking some fruit home.

Martine and I dallied a long time at her house that night, having several martinis each, before we finally decided we ought to go to Ed's house. He was not angry, though he had every right to be, considering our lateness. I think he had eased his waiting time with some alcoholic potion also. Anyway, we three soon got to feeling very amiable.

We sat around talking a while and then at my suggestion went out to pick some avocados. So, sometime around eleven o'clock I found myself high up in an avocado tree, peering into the darkness for fruit while clutching branches to avoid falling. There was no moonlight, so seeing the fruit was not easy. But I persevered and soon began dropping them while Ed and Martine scurried about in the leaves to pick them up. I remember feeling like one of those Malayan monkeys who have been trained to pick coconuts. It was hardly a professional way to gather avocados.

Perhaps the next most significant occurrence was my purchase of a house in Santa Monica. The place I'd formerly had in Pasadena had been bought as a family home, since I had then been married. It was quite large for a single man. So I sold out and looked for a smaller place nearer the ocean. Although I had got somewhat used to the smog, I did not like it, and I figured if I was going to be around southern California a while, I had better locate somewhere where there was the least amount. That had to be the west side. Also I like to think that a part of my motivation for moving was a longing for my own avocado tree. Although the Pasadena house had much shrubbery, including two large olive trees, there were no avocado bearers. In any event, when I saw a little Spanish style stucco house in Santa Monica, I was greatly interested. And when I saw that a fully mature avocado tree was flourishing in the back yard, I had to have it.

I bought the place and moved in. Soon I was tending my very own avocado tree, in retrospect fairly badly. I really knew nothing about the cultivation of this tropical evergreen. What experience I had had was with deciduous fruit trees in the northern zone of the U.S. I had never lived long enough at one time in the tropics to get a tree through to maturity. I asked around and discovered that the local gardeners and nurserymen really didn't know much about the cultivation of avocado trees either. Their expertise was for shrubs and plants for urban yards. As I recall, I over-watered and over-trimmed my tree. But despite the mistreatment, it produced fruit, though I'm sure considerably less than it would have if it had been treated properly.

So I lived happily in the little Santa Monica house, tending my avocado tree and garden, bicycling, teaching anthropology and leading a fairly active dating life. And then like quite a few other Californians (which I fully considered myself by then), I married again. My new wife also liked to eat avocados though I must say that now after some fifteen years in avocado land, that I have found few ladies who did not appreciate the fruit. In fact, I could easily recommend it as a romantic offering to replace flowers. But as I indicated before, southern California is also famous (infamous), for the brittleness of its marriages. Nevada may be where people from out of town go for quick divorces but California is the place where they most often develop conditions to want them. All that is needed are "irreconciliable differences," which after a few years of married life, my wife decided existed between her and me. She initiated divorce proceedings, and after a considerable amount of pain, we went our separate ways.

An important side effect for me was economic. After considerable amounts paid through court orders and property negotiations, I came out quite a bit better off than when I went in. What to do, since in the wings of this drama was the Internal Revenue Service? I was taking in too much money, and my deductions had dropped to little more than the $1,000 personal exemption. I was very likely to get zapped by the guys who study the Long Form. I had heard about tax shelters for people who made more money than they could use. This hadn't really been a problem before when I had been married. But as a divorced academic male, I was joining a group made up of doctors, dentists, successful lawyers and businessmen, divorced or still married.

Perhaps almost without conscious intent, my thoughts turned again to avocados. Those delicious fruits had been one of my continuing delights since coming to California, and I had long heard of the avocado grove as a tax shelter. I located a knowledgeable realtor and within a very short time was the proud owner of a 5+ acre spread of Haas-Fuerte-Zutano avocados, plus some twenty kinds of other fruit. Also, as part of the package, I got a most comfortable house on top of a spectacular hill in San Diego's North County. I threw myself with great enthusiasm into the tending of my new pride, while continuing to teach anthropology and write.

It was frequently hard for me to believe that I had really become a gentleman farmer of the gift of the ancient Mexicans. My life in the autumn of my years seemed to have reached an apex. I had my own avocado grove, not to mention all the other fruit trees and garden, a most spectacular hilltop site in a climate which I had found the most ideal in the world, a most comfortable house (a friend called it a "house-house"), a teaching position in a subject I dearly loved, anthropology, and good health. Life was not treating me badly, I thought.

Then I met Mary Bradley on an airplane trip to visit my son.

She was an attractive airline stewardess in her early middle years who admired my long white hair. I was then sporting a Buffalo Bill look. She was the first flight attendant who had showed any interest in me, and I suppose she was fulfilling a general American male fantasy. Anyway, when I came out of the washroom, I stopped to talk to her and her friend. I described my avocado farm and ended up by presenting her and her friend with some of the avocados I was taking to my son. She agreed to see me when she was off duty.

We went out a couple of times in her home town of Houston and had a good time together. Then when my visit was over, I returned to my avocado grove in North County. Being an airline employee, she could get free flights to all the major cities, so a few weeks later she came to Los Angeles, where I met her and brought her down by car to my avocado farm. Though the vistas there are spectacular, to city people the location frequently appears isolated. She seemed nervous at first though after several hours became more relaxed. We ended up having a pleasant weekend together, hiking around, visiting some neighbors, picking avocados, gardening, swimming and lolling about in the jacuzzi. We also made love. Then came Monday and time for her to get her flight back to Houston. I prepared fancy avocado omelets and served them with a good bottle of Colombard. We dined in the brilliant morning sunlight on the deck which overlooked the avocado grove. The food was good, the wine and sunshine made us mellow, and we reminisced about our first weekend together.

It was about an hour and a half before we had to leave for the airport when the idea came to me. Probably it had long been lodged in my consciousness, perhaps since my reading of <u>The Golden Bough</u>, the old classic of magic and ritual. But never had the circumstances come together in my life more appropriately, I thought.

I reached over to fill her wine glass and our glances met. We lifted full glasses and took a sip each. I was quite warm and had a little buzz on. She looked at me quizzically. "Now, what profound professorial thought is going on in that white head of yours?"

Despite my relaxed state, I hesitated. It was a bizarre idea, even for me. Finally I answered. "It's not very profound and I think you couldn't guess in a thousand years."

That, of course, piqued her curiosity. "Why don't you tell me?"

Then I thought why not. If there ever was a proper time, this had to be it. Still I hesitated, then, "It's anthropological."

She smiled, encouragingly it seemed to me, saying, "What does that mean?"

I went on. "It's about sex also."

Her impish grin widened. "That doesn't surprise me."

"You really want to know?"

"Sure, try me."

I took a deep breath. "Well, you know I told you I seemed to have been linked to avocados in some inextricable fashion ever since I ate the first one. And so to have got my own grove was the fulfillment of what I now visualize as an inchoate dream. On the other hand, I have been a dedicated anthropologist even longer, especially because of the great variety of exotic customs I have read about or observed. Well, one type which intrigued me from the first was the fertility ritual. It is a form of sympathetic magic as described in considerable detail by the early British anthropologist, Sir James Frazer."

She was listening closely. Although I had had no indication before that she was greatly interested in intellectual topics apart from those concerned with the development of self, she did read. I figured that she had never read anything in anthropology however. But for that matter how many other Americans had?

Anyway, she showed enough interest to keep me going. "As best I recall, the classic fertility ritual in peasant Europe as described by Frazer was an act of copulation between a village maiden and a youth in a field or grove under cultivation. Their sexual act was thought to bestow fertility to the ground and the plants that would grow on it. It is a magical idea by which like produces like."

I was pleased that she was taking it lightly, though I was not overly surprised. I had learned enough about her by then to know that she had had a variety of sexual experiences in her lifetime, and as an airline stewardess, she had been trained to deal with the unexpected. She was grinning, and with a hint of comprehension on her face, as if she was getting an idea where I was going.

I was still reluctant to come out with it totally, so continued a little longer in professorial fashion. "As best I recall, the old classic fertility rituals were most often done in relation to grain." Then I made a little transition. "And of course grain is not grown in these dry hills." I grinned then, particularly at the obviousness of my statement. "We specialize in avocados as you know." I took a deep breath and went on. "But interestingly enough, avocados have been regarded as having a sexual aspect."

Some years before, for a lecture on language, I had looked up the meaning of that buttery, green fruit and found out that the English word derives from the Spanish "'aguacuate." And as is typical in word borrowing, the Spanish word was modified by English speakers to fit the sound rules of their own language. But then I learned that the Spanish word had come from Nahuatl, the language of the Aztecs, as "aguacuatl." It too had been modified by the early Spanish to fit the sound rules of their language. But what was even more interesting was the meaning of the word in Nahuatl. It meant 'testicle fruit,' from its appearance of course. According to the dictionary, the specific variety from which the name was derived hung in pairs, one lower than the other. Her grin widened. "Hmm, interesting. So that's what anthropology professors do. Doesn't sound so professorial to me though."

I felt more comfortable then. "Well, I don't think all anthropology professors do these kinds of things, but we are known to be an unconventional lot." She smiled and I went on, "Well, it occurred to me that since I now have my own avocado grove, and since I am a dedicated anthropologist, and since you are here with me, it would be an interesting idea to perform a fertility ritual among the avocado trees."

The smile remained, even though she looked at me with what I thought was some surprise. Then she laughed lightly, "I should have guessed it would be something like that, coming from the only anthropology professor/avocado farmer I've ever known."

I still couldn't quite make out what her reaction was and continued to feel hesitant. I thought for a moment I had gone too far, so quickly said, "Of course, I should have thought of it earlier. There's so little time now before you have to leave to catch your plane."

But then to my considerable surprise she got up, keeping her wine glass in her hand. "I'm game. Shall we go?" I carried the blanket and she carried the wine bottle as we headed into the grove.

The next day I went alone to the tree and carved our initials in the bark. I had never performed that little magical act in my youth. It wasn't that I believed in the efficacy of such symbolism but I felt something special ought to be done to mark a classic ritual as described by Sir James Frazer. After all, how often had such been acted out by any anthropologist?

Then I dug a hole at the base of the tree and buried a small packet which contained a garment of hers and mine, wrapped in aluminum-foil. That had been a thought which came to me as I was driving her to the airport, a modern magical touch. I would leave a time capsule to retain and disseminate the force of fertility throughout the grove in the years to come.

It is now many years later and Mary Bradley has long gone her own way. It turned out that she was bisexual and while seeing me, decided she preferred women. Even the avocado grove is gone, due to economic reasons. However, all the other trees at Hilltop continue to produce very well. I tell myself that Sir James would be happy.

16

Us and Them

Probably the most constant attitude among life forms is concentration on self. This derives, of course, from the need for continuation. Everything alive has a built-in need to continue its kind, and the first prerequisite is for the individual to go on, at least for awhile. Wanting to die is not usual in nature, at least until each creature has lived out its life cycle, which always includes some means to make the species continue. Thus, salmon struggle throughout their life cycle to continue their individual selves. But once the female has laid its eggs, and the male has fertilized them, the adults literally give up their worn-out bodies and are consumed. The males of many kinds of herd animals, such as elk or kangaroo, wear themselves out struggling to maintain herd mastery in order to copulate with the maximum number of females. They hang on a few years as alpha male and then are replaced by younger, stronger animals. Even plants try to continue despite all adversities until the seed is developed. Stress a plant with drought and the last thing it will do is to produce a few good seeds. So, the individual concentrates on itself until the future of the next generation is ensured, after which it may or may not struggle to continue.

This may sound very obvious. And although I think the fundamental biological process is fairly self-evident, the ramifications, and especially those which are not biological, are very great. Their consequences have been enormous in man's history.

This is the matter of centrism, that life forms concentrate primarily on their own individuality, their centeredness. I suspect that such centeredness has been the cause of more human problems (and achievements), and less understood than most other forces that affect the talking biped. So, how does centeredness enter the human quotient? First off, man is no different than most other life forms in that the individual is primarily concerned with his own existence. Each wants to survive as long and as comfortably as possible. We call it egocentrism, or the concern of each person with his own psycho-center. A person wants to survive and achieve the gratifications his culture has taught him to desire.

Human egocentrism is different from that of other animals only in that the goals of the individual are those he has been taught rather than him being genetically programmed. Thus, while the male elk tries to eat as much grass as possible in order to maintain his physical capability to gain mastery over a large number of females, the human male may attempt to control his eating in order to get a highly paid job or to stave off a heart attack, the number of females to be obtained being secondary. The human male may even limit the number of his impregnations, the biological achievement being secondary to the cultural achievements of having money or power. But apart from the specificity of cultural goals, human egocentrism is very similar to that of non-humans.

Obviously, this kind of centeredness has great value to the group. If individuals did not try to maintain themselves and achieve the cultural goals, the group would lose its elan, its drive to persist. There have been descriptions of such occurrences. An Africanist, Colin Turnbull, described one such group, the Ik. Probably most American Indians, and other ex-tribals, have gone into such a state when they were being overwhelmed by whites. What was the use of reproducing one's own kind when the known way of life had little likelihood of continuing?

In any event, without a drive to persist in a world of competition, the group would certainly be in jeopardy at the very least. It is surprising therefore that we hear many criticisms of egocentrism. "You're just being egocentric! Or "That's very egocentric."

Jessica told me that when she was growing up, one of the commonest criticisms she got from her mother when she was trying to satisfy her own needs was "Me, me, me, me."

In such instances the individual is being criticized for not taking on super-individual responsibilities. The individual is making choices to fulfill her/his own needs by neglecting some group, the family, clan, ethnic group, nation, etc.

However, egocentric drives must have some positive consequences, in some cultures more than in others. Westerners, and particularly Americans, are thought to value individualism highly. Asians are thought of as being more motivated by group needs. The worker in an American factory is supposed to be concerned with getting the maximum benefit for himself directly, while the Japanese worker is expected to be working primarily for the "company."

There have been several explanations for the emphasis on individualism among Westerners. The one that seems most reasonable to me, though hardly provable, is that Westerners have been in the forefront of political and scientific expansion and change in the last 400 years. They stressed progress and the future, not tradition and the past. Thus a prime role has existed for the innovator, the one who would change things. And it is certainly true that change begins with individuals, the inventors. Furthermore, those who spread new ideas must be those who are going most against the inertia of tradition. That is typical of the inventor, scientist, explorer, entrepreneur and diffuser. That this was thought to be particularly true of Americans has also been explained as a need for rugged individualists on the frontier and in a rapidly expanding economy in the 19th century. But whatever the cause, the rugged individualist has remained a type much favored in American culture right through the 20th century. He is a prime type in American literature, exemplified by Henry David Thoreau and Ayn Rand. These were people who advocated doing one's own thing, no matter what. The most famous type of hero from Hollywood and television has been the western tough guy, exemplified best by John Wayne and Clint Eastwood. These were men who through sheer determination, and a fast draw, brought law and order to messy situations. The same type has been featured in police work, the tough individualist who solves the crime on his own.

Industry, commerce and politics have had their share of individualists. The robber barons of the 19th century were men who did their own thing, letting the chips fall where they would. Up to very recent times Americans have produced entrepreneurs who gained great riches by using other peoples' money, and taking wildly imaginative financial gambles, right up to the point of bankruptcy. Such were the men who produced the savings and loan scandal of the late 20th century. Some of the most flagrant such individuals who blatantly flaunted their wealth and individuality were Donald Trump and Charles Keating. The most telling observation is that the American economic system provided optimum conditions for such action. Individualism was to be highly rewarded.

And so am I saying that individualism is all negative? No, not hardly. It is obvious that a viable culture needs some egocentrics, those who will be in the forefront of whatever happens new. Someone has got to come up with new ideas if things are going to change. And if things do not change, the unchanging cultures will be absorbed by cultures which do. Because if there is one overall truism in life and culture, it is that things change. So there is a secure place for the inventors and diffusers of new ideas. Gautama, Jesus, Plato, Socrates, Einstein and Fermi were the front-runners of their cultures. And all had to be egocentrics to some degree.

So what's the big deal? Why not simply quit at egocentrism? Let the world be run by rugged individualists? Very simple, which is that while egocentrism is the basis of animal existence, both biologically and culturally, it is superseded by higher levels of centrism. In fact, the emphasis on individualism (egocentrism) in the West, and particularly in America, may be one of its most serious social deficiencies. Cultures elsewhere, and particularly in East Asia, have played down the importance of the individual while emphasizing the group. One works first for the group and then for the individual. Admittedly the fundamental unit is the individual. He/she/it tries to get what seems best for the self. But in the growing up process it has already been made clear that the individual must help fulfill needs of others. These are the successively larger social groups, the basis of existence for all social animals.

There are many animals which are not social--insects, reptiles, some fish and mammals. Among them rugged individualism can reign supreme. The only constraint on behavior of the shark or bobcat is what is inscribed in the genetic code. However, even these creatures have to cooperate on one specific occasion. They must have sex as a pair. Then if impregnation takes place, and they are not all killed off, the group will continue.

And what does such coupling have to do with the talking biped? Like other animals, including the non-social ones, man has got to get together with his mate. But as we all know, it doesn't stop there. Man goes on to form a larger group, known as the nuclear family, consisting of children and mate. This group stays together and cooperates in several ways and has continuity beyond the individual. There is a theory that the nuclear family was the original social invention and became the basis for all the more elaborate social groups that evolved. Anyway, man did go on to form many such, which became what we know of as kinship.

And how does the family or larger kin group take on an important role for man's way? In the same way that a culture can emphasize the ego, it can also emphasize family or kin. Then individuals emphasize their family over all other social groups or individuals. They say, "There's only one thing you can count on, the family."

Individuals see their cooperation with other family members or relatives as more important than cooperation with any larger group or even their own egocentric needs. This is widespread. There are large cultural groups which emphasize the significance of the family over all else. Some such that quickly come to mind are cultures from the Middle East, India, Pakistan, the Far East, and Latin America. Whatever else happens in such cultures, the general belief is that one can count on the family.

Family relations seem to have diminished where industrial urbanization has evolved. Thus, in the West the nuclear family predominates and the greater family is thought of as a problem. Even so, the nuclear family has weakened. Divorce and the temporary family have become common in urban, industrial society. Also couples have less interest in raising children. They establish temporary relationships with one another on the basis of emotional feeling (love) rather than kinship connections. The number of children per family has decreased so much that nowadays in the urbanized, industrialized West there are countries with negative rates of reproduction.

Is kin-centrism either good or bad? The final score is of course not in, though kin-centrism seems to be a condition which is being phased out naturally in urban, industrial society. Evidently the most that can be expected, as the world continues to industrialize, is to slow up the process. But that still might be critical for cultural competition between national cultures. The greater family orientation of the Far Eastern cultures seems to give them an advantage in the sense that they maintain more responsibility for education and economic cooperation.

On the other hand, family orientation probably decreases the importance of the role of innovator since these must be individualistic to a degree. Overall however, the kin orientation of the Far East seems to be more beneficial than the individualism of the West. There is little doubt that even in the Far East many of the responsibilities of the family have been lost as urbanization and industrialization have increased. But still the kin orientation of cultures like Japan, Korea and Taiwan remain stronger than in Western Europe.

Furthermore, kin orientation has been extended in the Far East. In a broad sense "the company" or "the team" in Japan, and to a lesser extent in other East Asian countries, has become a super-family. An individual is supposed to work for the group and the group will take care of his interests throughout his lifetime. The idea of the super-family is much weaker in the West, where in productive activities the individual is supposed to look out for himself. At one time the trade union served this purpose partially, but this group too has become greatly weakened in the West. Thus one who is dissatisfied with work conditions or pay is considered to be within his rights to seek another job. As a consequence most people have had several jobs in a lifetime. Apart from extra jobs, I have had work in seven institutions during my work years. All were taken because I was dissatisfied with my previous job. In social science jargon we call this horizontal mobility, and it indicates a primacy in fulfilling individualistic needs. The Japanese, on the other hand, specialize in what we call vertical mobility, movement within the company. One is expected to stay in the same place and move up through ability or longevity. Of course most people only go so high, and some go down, but this is true in the Western workplace also. However, jumping around from one company to another is looked upon with disfavor because it indicates a higher regard for the individual's needs than for the needs of the group. The dedication to the group seems to have made a more efficient productive unit of the Japanese company, as it did of earlier pseudo-kinship groups.

Men organize themselves into ever larger groups as societies become more complex, well beyond kinship. Any ideological core and set of customs shared by a group is enough to make them act centrically. Because of shared beliefs and behavior, and frequently some mythological beliefs, a group of men will insist on their own primacy and attempt to dominate others, to impress their own ways on them.

One of the most spectacular of this kind of group is that based on religious belief. Though there have been many religious systems, particularly in tribal societies, which had considerable degree of tolerance toward others, almost all of even the smallest groups believed their own way was best. But as groups evolved with greater clout and numbers, intolerance grew. By the time we get to historical times, intolerance from centrism has become more and more common. One of the earliest religious systems we know of which maintained strong self-righteous centrism was Judaism, even though the society did not have the ability to prevail politically against any of the great powers through most of its history. The Jews considered themselves the chosen people of god, to them the center of the universe. A more centric attitude would be hard to imagine, even if it was typical of most religious mythologies. There was no evidence in this world that the Jews were chosen, since they were unable to maintain even their home territory through most of their history. However, religious ethnocentrism, which such was, is rarely subject to disproofs. And the Jews' belief in their own rightness was so strong that no matter what indignities and atrocities they suffered, they maintained a belief that their way was the right one until they were at last able to re-establish their own state.

And from this seed, two other great systems emerged which were equally self-centered, Christianity and Islam. However, they contributed a new element to history. Judaism was self-satisfied in its centeredness. Jews believed that their religious system was the best in the world, but for the people called Jews only. There was no need to spread this blessing to others. It was enough for them to guarantee the salvation of their own people. This of course was the reason that Judaism never grew as big as either of its daughter religions. Christianity and Islam were different in this respect. Both set out early to convert everybody else. All peoples could be Christians or Muslims if they accepted the basic teachings and followed a few simple rituals. The age of conversion had begun. We know what happened. Millions of people adopted the central beliefs of these offshoots of Judaism, and apart from Buddhism, the world came to be filled primarily with their followers. The world of man became religio-centric as it had before been egocentric and kin-centric.

The shared beliefs of the followers of these religions brought about massive changes in world cultures. For instance, Islam carried from Judaism a belief in the abomination of graven images. One could not artistically depict images of spirits, and by extension the images of ordinary men. As a consequence massive destruction was wreaked on the images of others from West Africa to India.

Buddhism was the one other converting religion which succeeded on a worldwide scale. Gautama's message was for all peoples, as were those of Jesus and Mohammed. Furthermore, like the two from the Middle East, most followers of Buddhism were converts from other religions. In fact, Buddhism was like Christianity in being very successful elsewhere while it almost died out in the land of its founder.

Buddhism was probably less centric than the religions of the Middle East. It accepted and incorporated many local beliefs and practices rather than damning them. There was nothing like the belief in the abomination of graven images of other peoples' spirit beings. One did not need to abandon belief in Lao-tse, Confucius or Shinto to be a practicing Buddhist. So it seems that centeredness of belief in religion or other belief systems is a force that can push men toward imposing their beliefs on others, either through force or by less direct pressure. Furthermore, it can be a powerful force toward unification and dedication, which facilitates conversion. There is little evidence that there is anything superior about any of the converting religions except their self-righteousness and aggressiveness to convert.

It has been frequently claimed that the idea of one god is a superior notion, but I fail to see why. How can a belief in a single anthropomorphic deity be any loftier than a belief in the spiritualization of all the forces of life, what we call nature worship?

Anyway, a religion or any other powerful idea system can cause vast changes through its centric force.

Probably the second most widespread kind of centrism is simply ethnocentrism, the belief that one's own particular culture or subculture is the best. The idea of ethnocentrism has become widely known in my lifetime and generally has a negative connotation. To get a population to be against ethnocentrism is easier said than done, however, and may not be advisable in any event. It is a force we deal with in our everyday lives and is a consequence of the continuous separation and amalgamation of peoples in the age of cities. Once the city emerged as the primary human social unit, many groups were conquered and moved from place to place, while other groups and individuals moved voluntarily toward what they thought would be better conditions. Thus, the city-state became a hotbed of ethnicity, made up of groups who had different customs, and from hither and yon. Furthermore, there was usually some group which achieved dominance. Its members looked down on those they controlled. The downtrodden, however, usually tried to maintain some integrity, and this frequently came to be their language and culture. And when the distinction between a group and others was great enough, usually as a minimum having a separate language, an ethnic group was born. Sometimes however, even a separate language was missing. The American blacks ordinarily consider themselves, and are considered by others, to be an ethnic group, even though they speak some dialect of the language where their ancestors were brought as slaves. Also they have few customs from their own ancestry.

The Jews are another case in point. Considered an ethnic group in Europe, where most found themselves after the Diaspora, the Jews learned to speak the languages of their immigration. Even their own "ethnic" language, Yiddish, is a second stage European tongue, carried from one country to another as the Jews migrated. It was originally a dialect of German, learned in the ghettoes of Germany, then carried to the ghettoes of Eastern Europe and finally to America and other countries of migration. However, throughout their long history of dispersion, the Jewish religious leaders used Hebrew. So a new language, Ivrit, was created by the founders of the new state of Israel. Of course, by then the Jews were no longer an ethnic minority since they had become dominant in their own country.

Anyway, the most frequent occurrence with ethnic groups is that they try to hang on to their old language and culture for one or two generations after being taken over. Then the new generation takes on the culture of the dominants. The Jews and a few others have been different in this regard. Although they always took on much from the dominant cultures, they at least tried to maintain their identity. And as long as some core of customary behavior remains, particularly the language, there is a force for ethnocentrism, of a belief that there is something special about one's own ethnicity.

The classic kind of ethnocentrism, which one is supposed to be against nowadays, is that of the dominant group. In America this is Anglo-Americanism. And as would be expected, the people of dominant sub-cultures are generally more convinced than anyone else that their culture is best. After all, they have already succeeded in becoming dominant, so it should seem reasonable that the ones they have taken over should become like them.

So is ethnocentrism as bad as the current attitude toward it implies? Isn't there something worthwhile in seeing one's own culture as the center, as somehow better than that of others? Indeed, I am sure there is something positive about the ethnocentrism of dominants. After all, it takes some belief in rightness for a people to go on to try to achieve. All the great civilizations were so. The favored citizen of ancient Greece and Rome certainly believed his culture was wonderful, as did the Inca and Aztec. The Chinese in their days of glory saw themselves as citizens of "the middle kingdom" which had no need for the gewgaws of outsiders. The Japanese tried to close off their country to keep outsiders away until Commodore Perry, loaded with ethnocentrism, came to forcefully introduce them to civilization. And now that the Japanese have become successful in producing and selling products originated in the West, they again see their way as superior, this time as internationalists. In short, ethnocentrism has clearly been part and parcel of great cultural achievements.

So why is there so much criticism of the attitude nowadays? Almost all the criticism I know of has been against dominant ethnics imposing their way on the ethnic groups they control. Thus, in America Anglo-Americans have been taken to task through legislation and in the media for discrimination against minority ethnics, particularly Blacks, Indians, Latinos, East Asians and Jews. I'm sure the discrimination is real, but it is also straight out of the national founding. The U.S. was created on the basis of ethnocentrism, as were practically all cultures which became dominant. The initial policy of the Europeans who established themselves on the Atlantic shore was to displace Indians, then Spaniards, French and English, while using Blacks for agricultural production. This kind of policy was followed by practically all other nation states as they became dominant.

There was one major change which now creates some conflict. In the last 200 years a belief in human rights for all evolved, especially among Europeans. This was of course no problem while Euromen were in the process of "taking over" and using minority people. However, once the minority ethnics were subdued, a change took place. Thus, we see a considerable change in the attitude of Anglo-Americans toward Indians. While the Whites were struggling for dominance, there was "no good Indian but a dead Indian." But since the land of the Indians has been for all intents taken over, and their cultures destroyed, Indians are accepted as petitioners for rights. And especially the Black descendants of slaves are now considered to have basic rights.

This is the new liberalism, more or less shaped by Europeans. One can hardly criticize it as right or wrong, except that it is clear that it came into being only after the question of dominance was settled.

Ethnocentrism can take strong forms. It has been popular since W.W.II to condemn Nazi Germany for its ethnocentrism against Jews and other "non-Aryans." Of course, Jews have been a focus of discrimination in Europe long before the Nazis came along, and for comprehensible, if not justifiable, reasons. They refused to assimilate, even if they took over many things in the local cultures. Furthermore, they became successful competitors to the dominants in finance. This combination has invariably led to ethnocentric discrimination. It challenges the dominants. Anyway, Nazi

Germany was only the most flagrant discriminator against the Jews. And since they lost the war, it was natural to blame them for an attitude that was widespread in Europe. Anyway, ethnocentrism can cause damages and even destruction of ethnic minorities in its most virulent form. And as long as a general value exists that all peoples, no matter their ethnic background, have inalienable rights, there will be a conflict between attempts to dominate and liberal treatment.

It might be pointed out that this idea of inalienable rights is a very recent one, not shared by most dominant cultures in the past, and not excluding that of the national founders. When they used the term, they were referring to male colonials (themselves), not Blacks or Indians or women. Also it is highly unlikely that the ancient Greeks spent much time worrying about the peoples they colonized or enslaved, and the Romans in their heyday certainly did not worry about the rights of the Germanic barbarians, as did not the conquering armies of the Inca and Aztecs. The Orientals were little different when their dominant ethnicities took over. So far as we can make out from their approved history, the Japanese suppressed and eliminated their native people, the Ainu, much as Anglo-Americans did their Indians. So the big question is the nature of the values of a culture, not its ethnocentric tendencies. These latter will push dominants toward ethnocentric control, which can lead to ethnocide. Perhaps the concept of inalienable rights can over-ride the force for control. However, this certainly remains to be seen. The Israelis, a people who had been on the other side for long, acted much like other dominants when they got in control. They discriminated against Arabs.

And so to another larger unit of centrism, the nation-state. It naturally tends to be ethnocentric because it is practically always controlled by a dominant ethnic group, which attempts to suppress and/or absorb other ethnicities. I am not here talking about physical differences. Although physical appearance may be useful as an indicator, ethnocentrism and nation-centrism is at base a matter of culture, no matter how much we talk about racism.

What dominants really object to is that other ethnic groups have different customs. And as was learned long ago in anthropology, any people is capable of learning any customs, and thus culture.

So nations are ordinarily ethnocentric. Their people are taught that their system is better than others in some important respects. If it is apparent that one nation is not technically or militarily superior, the teaching will be that they are superior in artistic achievements, religious beliefs, social services, or whatever. And as is well known, nations get into disagreements about their differences which often turn into wars. This, along with territorialism, is probably the main cause for war making. Prior to the rise of the city-state, people also went to war, though admittedly on a smaller scale. And since the pre-urban social unit was usually a homogeneous ethnicity, the force for war was ethnocentrism. All that happened differently with the rise of the nation is that the larger social unit was tied to the dominant ethnicity as a model and warfare was on a grander scale. There is little doubt that most Amerindian groups fought one another, usually for territory, but they did this as ethnic groups. The Dakota fought the Blackfoot while the Navaho raided the Pueblos. When Pax Americana was established, the remnants of the Dakota, Blackfoot, Navaho and Pueblos were taken into the Anglo-American army to fight the Germans and Japanese.

As a soldier in WWII one of my best friends was a Winnebago Indian, who I and my Euro-buddies referred to in our unabashed ethnocentrism as "chief." Needless to say, I had hardly heard of anthropology then and the "chief" had long learned how to be a "friendly Indian."

A change in social organization took place with the rise of the nation state. The leaders of nations discovered many new ways to exploit the common folk, and one such was the conscript army. In pre-urban societies fighting was almost always a voluntary affair. Moreover, there was no salary. In the new national societies, it was discovered that with proper manipulation of the masses, the leaders could demand military duty for patriotic reasons. One could then pit the peasants (read "working class") on one side against the peasants on the other. Also governments could set up much of the productive capacity of the workers to make war machines, the salaries of soldiers and cost of war machines to be paid for by the taxes of the workers. Warfare could become serious business.

And so long as nations remain independent centric groups, there is a likelihood that wars will erupt between them. The wars will be controlled by the dominants through their leaders, using the minorities and lower classes as soldiers of the line. And as with other forms of centrism, there is nothing good or bad about this except insofar as it violates the value system. If a people has a value that each life is sacred, then obviously they will want to keep men from coming home in body bags, even if they are of the lower classes. But if warfare is seen as a legitimate way to settle differences and achieve territorial expansion or economic gains, as in 18th and 19th century Britain, then warfare and body bags are okay.

It must be remembered also that this is no matter of eliminating mankind as a species. Despite the steadily increasing ferocity of wars during the era of the city, there has been a constant increase in population, until in the 20th century population density is probably a much more serious problem for mankind than warfare. Warfare, even on a grand scale, cuts down the population no more than one generation, and that even in only the worst ravaged countries.

And what about super-national centrism? Men have long thought of alliances, either for protection or aggression, or both. In recent times, in the era of human rights, men have claimed they might eliminate or control warfare through such means. The two best known organizations for such have been the League of Nations and the United Nations. Both had enough varied membership that unadulterated ethnocentrism or nation centrism was difficult. But too many different cultures were involved for easy cooperation. So as we all know that for stopping war, or even decreasing it, the League of Nations was a complete failure and the United Nations almost so. Smaller wars have been constant and it is probable that another major war has not taken place since WWII because too many powers have got too many very deadly weapons, particularly the nuclear bomb. Anyway, members of the United Nations continue to be mildly tolerant, but depending primarily on the values of the West. Discrimination is not too likely between most of them also because they are spread around the world too much. Most conflicting customs are never put to the test. It doesn't matter if one eats dog, or horse, or beef, or is a vegetarian, if each lives in his own milieu where most everyone shares the same customs.

A kind of centrism on a grand scale which has had more effect has been that of civilizations. These are often many nations bound together by a dominant religion. Perhaps the two primary ones of history have been Western Europe and Islam. From the ashes of the Roman Empire, which was basically an ethno-nation-centric combination, a powerful new entity, Western European, made up of many national units and ethnicities, arose. It developed a core which all member nations shared -- the religion Christianity, the idea system science, a strong value for change, particularly in technology, and a different political structure. Also there have been special artistic traditions, both visual and aural. Almost all nations used Indo-European tongues and the basic physical stock was light skinned with much body hair.

There was a lot of interaction between the nations and diffusion of new customs was rapid. Thus a commonness of culture emerged. Western Europeans came to conceive of themselves as a unit, so much so that when their day of global expansion came, they frequently cooperated, something they practically never did with the natives they encountered. Thus, the Spanish and Portuguese more or less amicably subdivided Latin America, while the English, French, German, Portuguese and Americans cooperated in cutting up China. Also Western Europeans bought and sold other peoples' territories among themselves, paying no attention to the residents of those places (the Americas, Pacific Islands). This is not to say that they always worked in harmony, particularly when they were taking over native territory. But even when they fought one another, they did so as one civilized nation to another, rather than as a civilized nation to a barbarian one.

The other great civilization has been that of Islam, founded by the followers of Mohammed. The core was specifically the religion with some Arabic spin-off. Thus the language of religion was Arabic, in which the sacred literature was written, and in which ritual events were conducted whenever possible. A few other cultural characteristics were shared such as male dominance, and an emphasis on poetry and calligraphy. But the overwhelming core characteristic was the religion, Islam, and its influences. Thus a member of this great group felt an alliance with others who worshiped in the prescribed way, who made pilgrimage to Mecca and eschewed idolatry.

The centric passion was strong and became in the end one of

the primary means of creating a cooperative force among the followers, and particularly to resist the cultural and political onslaught of Western Europeans. If nothing else worked to create harmony among Muslims, they turned to their common faith.

Of course people of other religions, and particularly Christians before the age of secularization, also turned to their religion to evoke cooperative feelings, such as those during the Crusades. But this was never to such an extent as among Muslims, and particularly after secularization became endemic.

One might include Buddhism as a force for unity and centrism in East Asia, but it never produced a civilization where the members worked together because they were Buddhists. China, Japan, Korea and Vietnam, all became Buddhist nations of a sort, but during most of recent history they were at odds with one another, often at war.

And how can we judge centrism based on a common civilizational core? First, it explains much historical aggression and acculturation. Both Western European and Islamic civilizations have been quite expansive, and the world we know today is mostly a consequence. As usual, leaders of the nations that assumed leadership roles in their respective civilizations thought their own variety was best, and still do. Political leaders may phrase terms more carefully nowadays, but when tensions rise, action is normally taken on the basis of centrism, either national or by civilization, or a combination thereof.

But the old us-them point of view doesn't stop at civilizations or religions. It also takes in the major split which occurs in the species, that of gender. However vital it is to have male and female for the continuation of the species, this does not mean the two have to be treated as equals; and of course throughout history they never have been. It can be supposed that the first male anthropoid who stood up and started making tools became aware that he could lord it over his mate. He was bigger and more muscular, and at that stage of evolution that is what counted. Throughout the long era of gathering and hunting by women and men, the dominance of the male was kept in check, primarily because the female was so important as a gatherer of seeds, roots and berries. People may have liked meat more, but the reality was that the family and band kept going primarily by means of the plant products brought in by the female. Even so, the male took over the roles of power, to defend the territory, to keep other males away, and to cure the ill. So "machismo," or male centrism, continued in a milder form than later. A turning point seems to have been when plants and animals were domesticated. Because the female collector knew wild plants better, she probably became the original domesticator of plants. But not long after, the idea occurred that animals could be turned into slaves also, though some of the big ones were hard to handle. So who would be more appropriate than the muscle man who had killed the same kind of animals before? The male became the domesticator of large animals.

Then another discovery was made. Why not use the big animals to cultivate the newly domesticated plants of the females? A device was needed. The plow evolved, shortly to be followed by the wheel, the wagon and other devices that could be pulled by animals. And who was most likely to drive the animals? Man took over plant cultivation as well as all other activities which required a mule skinner. In the industrialized countries this lasted until the final days of family farming, and in the non-industrialized nations it still continues. And as is usual in society, those who control production also take over power and privileges. The authoritarian father was born, and woman became number two. The patriarchal family had evolved.

Men and women then saw their own sex as a group struggling to get, or maintain, privilege. Women were supposed to help by producing babies, tending the family and the house, while men would conduct affairs of state, go to war, vote, handle money and property. A clear us-them situation was created. Each treated affairs largely as a matter of gender. "A woman's place was in the home," "the man was the bread-winner," "women were overly concerned with feelings," "men were more logical," man was this, woman was that.

Now that the agrarian age is over in the industrialized cultures, the gender business has become somewhat less. One is supposed not to be sexist, which means of course that men should not discriminate against women, women can now vote and own property, and even divorce as freely as a man. But a residue of the old patriarchal centrism still remains in the industrialized countries. Women get less pay for equivalent jobs and most positions of power still go to men.

In the less developed (less industrialized) cultures, male centrism still reigns. The Latin male is "macho" and can play around with relative freedom. He can beat his wife if she does so. The Arab male conducts the affairs of state and commerce, and goes to war, while his wife is in seclusion even when she goes on the street. Even in highly industrialized Japan, where it happened more recently than in Europe, the world of power and privilege is that of the male. So overall, males and females still operate in an us-them pattern, the primary centrism being male.

But so far we have only been discussing differences among groups of men and women, but all of the same species, even if at different cultural levels. What about other forms of life? I propose that another form of centrism raises its head, homocentrism. Humankind at most levels of development came to the conclusion that people were clearly superior to other creatures. This may have been a little difficult at the earliest stages of evolution, when man was little more than a root and seed grubber, getting a little meat on occasion. A prominent theory nowadays is that during the earliest stage of evolution, which was the longest, man got most of his meat from leftovers of kills by the big predators. He was an eater of carrion to supplement his vegetal diet. Anyway the talking biped kept at making tools until there came a day when he was willing to stand up against the great cats, dogs and bears. And by the time Euroman discovered the multitudinous tribal people of the world, they had already developed myths that made them superior to other animals. The buffalo had been created for the Plains Indians' benefit and the seal for the Eskimo. True, many myths, such as those of the Northwest Coast, Eskimo, and Australian aboriginals described the other animals as being more closely related to man than was true of later, urban peoples. But there was no doubt in the myths of the hunters as to who was supposed to eat whom.

Then when man got urbanized and "civilized," new religions appeared that disenfranchised the other animals. No group did it more emphatically than the ancient Jews and their successors, the Christians and Muslims. All other animals had been created for man's benefit, primarily as meat. And in the end man became the slaughterer without parallel, of both domestic and wild creatures. He became such an efficient killer that he brought many varieties to near or complete extinction. He was the homocentric extraordinaire. Has this vast slaughter been an undesirable consequence of being homocentric only? I think most urban citizens in the West, and to a lesser extent elsewhere, wish their ancestors' depredations had not been so thorough. It would be nice to have a few passenger pigeons, Carolina parakeets, various Caribbean macaws and some dodos still around. It is popular in the West today to claim that we try to preserve threatened species so the children will have a chance to see them.

Much less often do the urbanized talk about the fact that there are quite a few species which are not on the endangered species list and which despite all of man's efforts to the contrary, continue to thrive. They have generally been categorized as pests and man continues to to try to eliminate them. There are many, but some of the most spectacular are the cockroach, fly, mosquito, rat, mouse, coyote, opossum and crow. They show no signs of dying out and if man ever gets to the point of blasting his own species out of existence, some of these creature will become the evolutionary inheritors.

In true homocentric fashion, man has always emphasized his own capabilities as the most superior overall. He has cited rational thought and tool making as paramount. We anthropologists would call this and its extensions "culture." But for sheer survivability and reproductive success, other qualities such as number of offspring, dietary inclusiveness, and smallness of size may be more significant. It must be remembered that man as a successful tool user has a history of only a few million years while the cockroach has been around for more than a hundred million. The other successful species also go back farther than man.

Anyway, for better or worse, man is stuck with culture for species competition and will have to follow that path as best he can. But he should harbor no illusions that the war of the species is over. Moreover, the odds are in favor of the others in the long haul.

Man's homocentrism has driven him even further. Once he came to the conclusion that he was far beyond other species, he decided that the earth was his place and the center of the universe. He became a geocentrist. And as we know, there was a mighty struggle between the forces of tradition and those of new ideas, particularly the ideas that the earth was only a small place in an enormous universe and man was only one among many species. This war is still not over, but enough of it has been won that a new biology based on change, and a new era of exploration based on the idea that the earth is only a very small part of an enormous universe, has evolved.

So what do we end up with in this endless pattern of centrism, this dividing up of the universe into an us and them? Each man, like all other creatures, is concerned with his own survivability, but he is also concerned with the continuation of his species. Thus the individual is egocentric, trying to survive whatever the odds. But man is also a social creature and gets considerable benefits from being a part of a group. So inevitably he sees the world as a matter of us (his group) and them (the other group). He works to make his group successful, ordinarily at the expense of the other group. He chooses his family, territorial group, ethnic group, race, sex, species and world as centric points to judge others. He is thus like all other creatures in a world of us and them.

So should man abandon all this centrism? I am not sure he should. He must keep enough at least to maintain a driving force for continuation. But I would hope he could learn enough of the marginality and temporariness of his position to cease the grossest forms and to take the others, whether rats or nations, seriously.

17

The Evolution of Presenting Behavior

In anthropology classes we frequently show a film on baboon behavior. Among other things, it shows how they try to be friendly. They run their fingers through each others' fur, which is called grooming. It is similar to caressing among humans. Also one will sometimes present his/her buttocks to the face of another. The presenter hopes thus to establish a relationship. And although these animals use such presenting to show friendliness in general, they specifically use it to attract another sexually. A female who wants to mate begins by holding her bottom up to a male's face. If he is interested and there are no interruptions, after a few preliminaries, they will copulate.

Most other monkeys also have such presenting behavior, as do other bisexual creatures. Birds, in particular, have elaborate show techniques. They puff up their throats, do wild dances, strut their stuff, and do water performances.

We anthropologists consider man to be an animal which shares many characteristics with other animals. He has the same biological drives, eating, drinking, sleeping, keeping warm and having sex. Also he is a social animal like a baboon or zebra and needs to show friendliness, to build relationships.

However, man seems to be unique in one important way, he has a culture. This comes from his special kind of learning, which depends on language. The other animals learn through direct experience or by imitating others. But they cannot learn very much this way. Man's way allows for a much greater variety of solutions. Thus, he can learn any kind of dance, a kind of presentation that someone demonstrates and explains. It is important to note the explanation factor. Learning by imitation alone limits the amount of new ideas a creature can get. A particular human can do a ballroom waltz, jitterbug, a tango, or all three. These are not "natural" to any individual or group. They are merely what has been taught. Man is the only animal which learns by means of verbal explanation, in addition to imitation and experience.

The other human needs are fulfilled in the same way. Therefore, it should come as no surprise that men have figured out many ways to present themselves to members of the opposite sex. Initially they concentrate on appearance. This is similar to animals and birds. However in the specifics of fulfilling this animal drive, men rely on cultural methods. They follow the current dictates of fashion, particularly of their age and sex group, in fixing their hair, skin, clothing styles and anything else that is visible. Skins have been painted, tattooed and scarified, while lips, noses and earlobes have been pierced and distended, and bones have been permanently reshaped by binding. The variety of clothing styles in different times and cultures is endless.

For the second impression, *Homo sapiens* most clearly departs from his cousin, the baboon. The absolutely human thing occurs when people start talking to one another. The ultimate in human presentation is with words. The gentleman wins "my fair lady" with "sweet talk," or "a line," and the lady of course does the same. This means of communication, language, is very complex. Almost everything in the visible universe, as well as many invisible things, has a word in some language. This generalization definitely includes all the behaviors and appearances of people. If a female baboon had a language, she could let it be known through the gossip network that she really goes for big, long-toothed types. A male baboon could let it be known that he favors females with supple fingers. Instead, baboons have to study each other until they spot the right type. This may be alright for baboons, who probably are less particular than people. But then again, the hairy primates do not need to consider their prospective date's occupation, salary, marital status, religion or ethnic background.

So how does this verbal presentation work? The archeological record tells us nothing in this regard. So we turn to pre-urban or tribal man on the assumption that his way of life is the closest we will ever know of how our primitive ancestors behaved. The ethnographers of the simple societies tell us that their people presented in a variety of ways. Some hunting and gathering peoples tried to control presenting through betrothal. The older people promised and gave away young girls for establishing alliances. The young girls were consulted very little, if at all. However in most tribal societies the young people could present themselves to each other to a much greater extent. They met at dances or other festivals or on paths outside the villages or encampments. Men gave presents, sang songs or played musical instruments. The flute was a favorite, partially I presume because it was small and highly portable, so the young fellow could hang on to it if he had to run. Anyway romantic notions are often reported as important in such presentations. The young people talked and arranged rendezvous. Such meetings could lead to premarital or extra-marital affairs, or to marriage itself, depending on the culture.

Then came the agrarian society, which was more complex and in which inheritance came to be important. The arranged marriage became the norm. In the classic arrangement, family elders sent word out through relatives or marriage brokers that they had a boy or girl ready. Both traditional Chinese and Indian society had such a procedure. In traditional Jewish culture there was a specific person, known as a "shadchan," who served as marriage broker. Negotiations were carried on between the intermediaries and/or parents. The couple was consulted to a minimal extent, depending on the degree of freedom allowed to young people in that society. In very strict societies, such as those of rural India or orthodox Muslim countries, the young couple would only get a look or a short visit under the sharp eyes of the elders. Almost all information and decision making would be in the hands of the older folk. In all cases, the goal of such pairing would be marriage. Dating would not exist.

Emigrants to new lands have faced particular problems, the most common being that the first to go were males without womenfolk. But by using the printed word, they frequently worked out systems for locating spouses in the country of origin. The Chinese and

European immigrants who came to America in the 19th and 20th centuries worked out such systems. Thus a Swedish farmer or miner would have an ad put in a Stockholm paper for a wife. Also native-born miners, farmers and ranchers in the American West put ads in Eastern newspapers for mail order brides. Although a photo might be included, the couple would have to make up their minds primarily on the basis of the verbal descriptions. But unlike the classic arranged marriage, the couple did it on their own.

Verbal presentation in modern urban society begins early. Girls and boys learn how to talk to each other. The young are primarily influenced by their age-mates to begin presenting themselves while they are in their early teens. Before they had got by in teasing one another. But the teasing stage must end, at least partially, if intimacy is to occur. So the couple has to learn to communicate, primarily with words.

After all these years, I still remember my first date vividly and even her name, though I never went out with her again. On my return from school one day, a group of girls caught up with me and urged Virginia Weber to ask me to take her to the upcoming scavenger hunt. That was a game young people used to play, in which couples would go out in the evening to collect items on a prepared list. The couple who got all the items first would get a prize. It was mainly a method to force shy couples to wander around in the evening together. My recollection of the evening is that it was extremely embarrassing, especially what conversation there was.

Boys and girls have to convince members of the opposite sex that they are desirable, which is not always easy. After all, before the communication stage, they had managed with different vocabularies. So one of the first things that has to be done is to learn the other's language. Thus, boys learn to talk some about clothes and social events while girls learn some about sports and mechanics. This may be a less difficult process than it was a few decades back. But there is probably still only a minority of girls who initially know anything about auto mechanics or boys who know much about clothes. But most of both sexes learn enough to get by in the dating game.

During the next few years, young people become increasingly active in seeking each other out. Probably the first places are their neighborhoods and schools. However, when they take their first jobs, more possibilities occur. They size each other up initially by appearance, but then comes the introductions and talk. With light conversation they try to tie in to each other's interests and activities. They continue until they decide they made a mistake; or they find each other interesting enough to try again. They make arrangements to meet and exchange telephone numbers. The interchange is almost entirely verbal.

The modern city presents some special problems, especially as people get older. In metropolitan areas, there are now extensive areas of apartment houses rather than single family homes. People know few of their neighbors. Also people move often in the modern city, from district to district, town to town, apartment to apartment and from one region to another. The native-born son/daughter is a scarce breed in the super-cities. In southern California a popular T-shirt proclaims "Native Californian." Also these strangers come from a wide range of occupations, ethnic backgrounds, religions, etc. They do not share a group of common beliefs and practices, like people in small communities. The chance of two people being naturally compatible is not great. So singles must look harder.

They continue to use the face-to-face methods they learned in their small towns or local neighborhoods. But these are frequently inadequate. So new methods of meeting people have evolved. Singles go to dances in commercial establishments, those under fifty to discos, those older to ballrooms. They also go skiing or on cruises, again commercially organized, to meet others. And there is the "bar scene." In all the large cities there are places where singles can gather, buy drinks, and meet. Generally, the same kind of presenting is followed as in traditional face-to face meetings. Individuals look each other over. Then by exchanging glances, they try to show their interest. The one who has enough courage introduces him/herself and the verbal presentation begins. Each must sift a lot of information quickly, primarily to decide how much is true. After all, the main social purpose of language is to influence others. So there is plenty of reason to emphasize the positive and neglect to mention the negative. Unfortunately, in the beginning there is rarely any other source of information about the potential date.

When singles had difficulties getting together in the past, they turned to advertising. So what would be more natural than to use such commercial channels for establishing relationships in a megalopolis? After all, urbanites live in a commercial world where most other needs are satisfied through marketing. A variety of listing services have emerged. The simplest is the commercial journal which can be purchased on the street corner. One of the most common in the West is the National Singles Register. It has articles on dating, entertainment and activities for singles. But the core is the presentation listings. An example: Very pretty, trim, well-proportioned, green-eyed blonde, excellent ballroom and disco dancer. Also enjoy theater, travel. Youthful 43. Seeking attractive male companion under 50, sharing interests, time. Photo please. West Los Angeles. S408.

One states one's qualifications and lists the characteristics one is interested in. A box number is included. The reader has to decide how much of a given listing is true. There is no way from the above to know whether the person is married or not, whether or not she is getting psychiatric counsel or how much she is in debt. And it must be assumed that no-one is going to put down negative characteristics.

Then there are the listing clubs. These are organizations which advertise for business. Interested parties have to contact the clubs and go through a screening. If they are accepted, they pay a membership fee and are put in the listing notices which are distributed to the members. Normally there is a code system to keep non-payers from using the service. Then the interested party has to call the answering service to get telephone numbers, giving his own to prove his membership.

Some organizations have developed sidelines. The one of which my wife and I were members also had social programs. These were encounter situations where people could meet personally at brunches, dinners, dances and trips, all for a fee of course. Presumably, these events were designed to compensate for the lack of face-to-face contact in the purely verbal listing.

A newer technique is a product of the electronic age. Instead of the informal talks of the Chinese or Jewish marriage broker, the computer is used for pre-selection. The characteristics of individuals are put into computer memories and the machine attempts to pick out compatible couples. References are forwarded to the selected persons, who are then on their own. Presumably this is an improvement, but only insofar as the individuals were honest in the screening interviews.

There is still the problem of a lack of visual image with the purely verbal descriptions, whether computerized or not. Anyone who has tried to use them comes quickly to the realization that comments like "young looking," "usually considered beautiful," and "handsome" are not absolutely reliable. After all, beauty or handsomeness is in the eye of the beholder. Besides, what man is going to advertise the fact that he is bald? While what woman with a shapely figure is going to admit that her face is scarred by acne?

So photo-dating. The listing service includes a photo of each individual. There is nothing new in this idea, since even in the 19th century ads a photo was sometimes included. Unfortunately, although this seems to be an improvement, a problem still exists. Contrary to the popular saying, pictures do lie, or at least skew the facts. Each photo may be a representation of the person the camera was pointed toward. But it is only one out of several. Naturally the individual selects the best one. And also one can fudge on time. If the listing people did not insist on a photo taken within the last year, why not offer one taken ten years before?

Individuals will be even less likely to put down negative personality traits. How could one possibly know that the "beautiful, sensual, shapely model" had a total inability to look another person in the eye? Or that the "handsome, athletic, well-to-do psychology professor" stuttered?

Another idea, why not make a visual recording of the individual in action along with a verbal description? Thus, the newest is the video record. Club members can look at the prospective's' performance before deciding. Theoretically this is an improvement. Certainly the individual will put his/her best presence forward, but still he will be unable to cover up the fact that he has a high pitched voice or that she has a tic. The method should favor those who have had some experience in displaying themselves.

The old standby still remains, the pure verbal listing service. This is natural since the ultimate presentation by people is with words. But what do they present? Let me describe what my wife and I learned after being members of such a service for about three years combined. Between us we dated almost one hundred people before we married one another. I should emphasize that this was before the AIDS Era.

The club operated in Los Angeles, New York and Miami. In Los Angeles, where we were members, it catered almost exclusively to professionals. Besides its social events, at one of which my wife and I met, it issued sheets of listings to members. A couple of examples: 123456. N. Hollywood, Ca. Dorie-35, white, 108 lbs., 5'4", red blond/blue (hair color/eyes), public relations. Loving compassionate, intuitive, intelligent, attractive, energetic, sensuous former stewardess, enjoys metaphysics, astrology, sailing, hiking, outdoors. Desires growth-stimulating non smoker, no dependents, 30-40, under 5'10, happily enjoying life.

Or: 654321. Glendora, Ca. George-40, black, 140 lbs. 5'10", black/brown (hair color/eyes), college professor. I enjoy sports, bowling, travel, dining, reading, outdoors. Would like to meet someone 35-45, 5'1" or over, warm, intelligent, attractive, preferably shapely. Should like children. Any race/ religion. Marriage a possibility. Prefer non-smoker.

The number is a code which one must use to get the telephone number and address of the advertiser. These are followed by the given name, age, "race," weight, height, hair/eye color and occupation. This information is required by the club, evidently on the assumption that it is the bare minimum necessary. This is probably true except for hair and eye color. It is doubtful that many people in modern American society screen out others because they do not like the color of their hair or eyes. These are unreliable in a verbal description anyway. Both are often changed by dying or wearing colored lenses by women. And it says nothing about quantity of hair. A woman can have no idea that the man, no matter how blond, has no more than a monk's fringe. The other physical characteristics, height, weight and age, must be important. "Race" probably is also, but since the club asked each person to specify his/her own, the terms were almost meaningless for anyone whose ancestry wasn't European. Black, white, and oriental were the most common. Thus, one couldn't know whether an oriental was of Japanese, Chinese, Filipino or East Indian ancestry.

Occupation is certainly important for presenting. After all, we live in the age of specialization so we type people by what they do. An early question of a new acquaintance is, "And what do you do?" If the man says he is an airline pilot, he is probably in, but if he says he is a television repairman, she is very likely to keep circulating.

Apart from our personal concerns, my wife and I were interested in how people presented themselves. What were they saying to one another in the listings? I thought of it as a way to know what values were important in this sub-section of society. Perhaps if we had listings of Chinese or Indian societies we would find them different. So we analyzed 1,000 listings and tried to put them into meaningful categories. I thought this might make a presentation profile.

The advertisers were not satisfied with the club requirements for describing looks. They put in much more evocative, detailed terms such as "lean, handsome, well-built, petite, attractive, slender, curvaceous, lithe, golden hair, long brown hair." It was apparent that they considered their looks very important. And since their appearance was not visible in the listings, they did their utmost to tell others about it.

Also they presented the details of their personalities. They claimed to be affectionate, sincere, sensitive and of generally pleasant disposition; and they wanted the same in others. Most of these people probably grew up in permissive families where they were taught that the primary goal in life was to fulfill their personality needs. They were from the "Spock/shrink" generation, reared with "let-do" methods. When later in life they had problems, many went to psychiatrists or counselors. They could hardly imagine being primarily concerned with social reality as a Chinese, Japanese, or Indian might.

They presented themselves as pleasure seekers, looking for like-minded others. All kinds of sports were okay, but those outdoors were preferable. Some preferred active sports such as tennis and bicycling. Others preferred easy ones such as walking. They liked to go on picnics and commune with nature. However they sometimes played sports inside, particularly bowling and ping-pong.

Quite a few (67%) claimed they liked to go to the theatre, concerts and movies. And when all else failed, they admitted they would watch television with their dates. Of course it is likely that they watched more television than they reported. This is not thought to be cool in the toney sections of society.

Apart from identifying their occupation, which was a listing requirement, they practically never said anything about work activities. These people were a part of the "fun" generation. It was hard to believe that they had anything to do with the famous Protestant work ethic, which supposedly has driven several generations of Americans. The closest we could discern to this ethic were requests for "successful" and "professional" males, usually by older females.

This does not mean that these people did not work hard. Dating and establishing relationships were for fun, however. Work must have existed for these people primarily as a means of providing more and better play.

Third, they presented themselves as having social interests. They claimed to be interested in dancing, parties, conversation, cooking and wine tasting. Females indicated more interest in these activities than did males. This fits the popular stereotype that females are more socially oriented than males. Also interest in social activities was presented more often by older people, particularly women in their 50's and 60's.

They claimed to be intelligent and interested in intellectual activities, particularly arts and higher education. Also they claimed to be interested in mind-stimulating games such as chess and bridge.

They were concerned with the kind and intensity of personal relationships, which was the most common term they used to describe contact with others. A smaller number were willing to list sex or sensuousness. All these terms imply sexual pairing. "To have a relationship" generally means to have regular sex, usually exclusive, with some meeting of the minds. Sensuousness and sex are feminine and masculine versions of pure sexuality. Females emphasized togetherness and companionship more often than males.

Do males emphasize sex more in their presenting because they have a stronger bio-drive? Some sociobiologists would say so. And is this why females emphasize what is more often slighted by males, companionship and togetherness?

Occupational status of males was important to females. Very rarely did females claim to be successful professionals, though they frequently asked for this in the males. Certainly there were more successful females in this group than the listings indicated. But our guess is that they did not present that fact. They did not believe that men valued economic success in females.

They did not present themselves as being highly interested in marriage. Out of the thousand, 189 mentioned this to be a possibility, while 47 claimed they were not interested. So about 15 percent were claiming they might consider such an arrangement. Probably more women were interested than claimed, but figured that this was not a high presentation factor. Although my wife and I married after meeting in this club, neither listed interest in doing so. Probably one's initial presentation would be more successful if sex, sensuousness or high economic status were emphasized, rather than interest in marriage.

A desire for raising children was presented even less frequently. Three percent stated they were in favor, while an equal proportion were totally against mixing schooling, baby sitters, discipline and teeth repair problems with their playtime.

Also very few people presented their religious affiliation (3%), though three times as many offered their astrological sign. For fun times they evidently thought that the practical use of astrology was far more useful than going to church on Sunday, or at least one could impress others more that way.

From this rough analysis, I will put together a profile of an ideal man and woman, drawn from the listings. But one must keep in mind that this is not the real, it is the ideal.

She: Beautiful, shapely stewardess/model/artist who is sensitive, affectionate, sensuous and has a pleasant disposition. Likes new experiences. Active in outdoor sports, particularly skiing, tennis and horseback riding. Also enjoys indoor recreation such as bowling and bridge. Likes movies, dining out, theatre, dancing, parties and togetherness. Seeks a meaningful relationship. Marriage and children a possibility, though not important. She is Sagittarius.

He: Good looking, tall, airline pilot/professor/actor who is sincere, kind and has a sense of humor and stable personality. Successful in his field. Very active in outdoor sports, particularly bicycling and sailing. Also into karate and jogging. Is an avid chess player and likes music of all kinds. Seeks sexual relationship for pleasure times. Not interested in marriage.

So there it is, a pure verbal presentation. It is a long way from the buttocks presentation of the baboon. But it is the human thing to do--to present with words. Furthermore, with the increase in anonymity of urban living, humankind will probably need to depend more and more on such direct verbal presentation.

18

On Being a Barnyard Watcher

When I opened the gate to the chicken run, two birds came running. Both of the Araucans, which lay green eggs, ran up to my feet, clucking the while. Where Araucans came from, and why they lay green eggs, I do not know. The word, Araucan, seems too similar to Araucanian, a large Indian group of South America, to be a coincidence. My best guess is that this chicken was bred out of general European stock in the region of the Araucanians, and has spread to other places as an odd-ball variety. They do lay quite a few eggs, even if green on the outside.

Anyway, what happened in my chicken yard could have taken place with almost any variety. All chickens, and other similar birds, have the same general reproductive behavior. That I know from more than fifty years of barnyard watching.

The two hens did a fancy dance around my feet, moving in close, then as I would bend over and reach down, they would move away, though not far. I finally reached the brown one and gave her a light flat-hand tap on the back. She squatted and half-spread her wings, making a little platform. I knew from long watching what would normally have happened next. A rooster which had chased her down, or had simply done a fast side step, would have grabbed her comb with his beak and then jumped on her back. Then as she moved her tail to one side, he would have pushed his penis into her multi-purpose sex organ, the cloaca, and have ejaculated. It was the basic act of reproduction, repeated with variations, among all complex species, all over the world, time after time.

However, from me the hen only got half the process. I picked her up and patted her back several times to mimic the rooster's tromping, and then pinched her comb and shook her head a few times to mimic the head holding procedure. When I dropped her, she shook her wings vigorously as hens do after that treatment. Then she strutted off stiff legged while clucking in the post-coital fashion of hens, as if to say to all, "Look what I did."

This procedure in my chicken run evolved by accident. I had raised seven chicks, primarily to help process vegetal leftovers from the kitchen and garden, but also to produce manure for the growing areas. I had not got a rooster at Jessica's request because of the noise. She liked to sleep late and felt that a rooster's crowing would disturb her early morning sleep. And a rooster is certainly noisy, even crowing in the middle of the night at times. So, though compared to battery-raised chickens, my hens had a paradisiacal life, one ingredient missing was that they had no feathered type to fuck them. And whether or not this contributed to their horniness, once they got used to me bringing their food and water, they elected me to be their surrogate rooster. They were lucky because I had spent so many years watching various creatures go through the same procedures and knew what the hens were after.

The earliest memory I have of animal fucking was at my grandparents' farm in Indiana in the 1930's. They did what was called mixed farming in those days, following simple though varied procedures. Thus, animals were expected to inseminate one another, sometimes with a little help. Artificial insemination had not yet arrived at the hill region of southern Indiana. They kept their chickens in an open flock, mostly hens though with enough roosters to service them. I observed the sex act there, undoubtedly for the first time.

Larger animals were more of a problem since they could be quite aggressive toward humans. The fact that their sex was rationed contributed to their ire. Farmers were particularly wary of bulls. But the cows needed to be serviced if they were to produce calves and milk. I saw one such impregnation when I was very young, perhaps nine or ten. As best I remember, the cow was moaning a long time before the bull was brought to her, he being controlled by a ring through the nose. It was a neighbor's bull, brought in by foot. My grandparents and uncles had no vehicle big enough to carry a bull. I have only a vague memory of the actual mounting, though I do remember that the men, the neighbor and my uncles, had great sport laughing and talking about the event. Perhaps men have done such since the age of domestication began.

My father and I kept rabbits and chickens in the city when I was growing up. Evidently this was legal then. There was always a rooster or so and I got to see the chickens coupling often. We bred rabbits with our own bucks so I observed them often also.

Apart from viewing these animals, I became familiar with cats and dogs, which were more numerous and rarely neutered in those days. Cats were a pain because they were more active at night and did the most god-awful meowing before going at it. When in heat, the females made long drawn-out moans, while the toms threatened one another with yowls. Not infrequently they got into screeching fights in their effort to drive off the competition. Dogs got into packs of 3-5 males when females came into heat. The males followed the ladies about, each trying to drive off the competition so as to mount the female. When one would succeed, he would get stuck to the female and the two would move about painfully. I learned later that at copulation a swelling occurs in a dog's penis which keeps the two stuck for a time, presumably for more effective sperm transfer. Anyway, the people of my city neighborhood then, having no sympathy for the plight of copulating dogs, and suffering under the prudishness of American morality, would run out in the street and throw water on the stuck dogs. The dogs would piteously try to pull apart. When one considers the prudishness of the time and the low regard for other animals, dousing them was a perfectly logical act. Apart from these animals, as a dedicated barnyard watcher by this time, I saw little else other than the copulation of a few species of city birds, sparrows and pigeons mostly. There was little exceptional about the sex habits of these creatures.

Instead, as a city boy, which I was despite my country

experiences, I shifted to humanity about this time, my early teens. One of the early uses of my newly acquired reading ability was to get sex books from other boys. I never bought any, but there were always plenty around. I exchanged them and stashed them away when not studying the pictures. I remember a dump outside town where I and a friend had a makeshift shack in which we had a cache of "dirty books." Many pleasurable hours were spent leafing through the sexy "big-little books" and others, fantasizing all the while.

Another new form of watching I learned from my contemporaries at this time was window peeping. The other boys would learn where there were windows with cracked shades or spaces, inside of which ladies sometimes undressed. When it got dark we would creep up close and watch. This was a scary game, however, because those inside frequently guessed that it was going on and would send someone out to catch us. Even so, I did it a few times, though without much success.

I feel sure most of my readers know that such watching was not a topic one shared with one's parents, or any other adults. Whatever sex information we had, we got from one another, or from seeing in nature what we could, human or other. The prudishness of our culture was so profound that sex, except for breeding animals, was not even talked about by the country men, and I presume women, even though they saw it every day among their animals, and arranged for it to happen for breeding purposes. Almost needless to say, adults did not discuss such things with children, and I suspect kept animal breeding and human copulation in two separate categories in their own minds. After all, Christianity had from the beginning demoted non-human creatures to a very subservient position, mainly to serve as human food. It is impossible for me to imagine Christians entertaining ideas of other animals and humans copulating together, an idea not at all uncommon in the thought systems of nature worshipers. So how could human love making, when mentioned at all, be discussed in the same breath as animal copulation. Having rejected Darwinism, and knowing little other biology, how could my parents have visualized their own species as having anything in common with chimpanzees or gorillas? So far as I remember, my mother never brought up anything about sex among any creatures, human or other. My father, who had grown up on a farm, called me into the bathroom for a "'talk" when I was in my late teens. He was so embarrassed that I tried to help him quit as soon as he gracefully could, thanking him for his help. By that time I had built up quite a little body of knowledge about sex from talking to the other boys and watching animals. A lot of it was inaccurate, of course.

But what about kids that grew up on farms and saw sex going on? Didn't they get more involved? In the first place, most being good Christians (my relatives were all Catholics), they had a tendency to keep animal and human behavior in separate categories in their minds. Still, it was obvious that the copulation of animals was like that of humans in general. But apart from the differences in anatomy, and thus positions in the sex act, animals obviously did not have the cultural overlay that humans had, of sex and all other behaviors. In particular, there was no indication that copulation needed to be done in private.

However, domestic animals were used by humans and so I suppose it was inevitable that some young farmers would think of using them sexually. And though I never observed any such acts directly, or even heard about any in particular, this was a part of male folklore when I was growing up in the city. Farm boys were supposed to be particularly fond of sheep, I suppose because this animal is close to humans in size, while still being fairly easy to handle. Also there was quite a bit of ribaldry about chickens. Young men of the working class, of which I was a part, not infrequently called one another "chicken-fucker." I always had difficulty visualizing how a human could copulate with a chicken but after seeing the Italian film, "Padre Padrone," and after my experience with the Araucans, I found it easier to understand. In "Padre Padrone" a group of Sardinian shepherd boys closed off the animal pen when their father left them alone and copulated with all the animals they could handle. The father in the meantime was copulating with his wife at the farmhouse.

Another spin-off from sex behavior on the farm was a repertory of jokes about bestiality. These were very common when I was growing up in the city and even into my middle years. Awareness of bestiality continued in male society at least through the era when humans were still living close to their animals and animals were becoming impregnated through actual copulation. I think most large animals are now artificially inseminated.

So there were a considerable number of conversational no-nos about sex during most of my life, and particularly between men and women. There still are of course, though less than in my early days. Which brings me to explain why I am writing about this topic. I am sure some readers will have decided that this is all "dirty stuff," far below their consideration. But there will be more now than before who will see my treatment as interesting, and perhaps even worthwhile.

After all, in the 70 plus years of my awareness of sex, we Westerners have gone through the sexual revolution. I am not sure it has been a real revolution, but I do believe that considerable change has occurred. There is less prudishness than there was when I was a boy. Now even "nice" girls can use four letter words on occasion. But even so, some bothered readers might ask, "What for? What's the big deal about sex? Who needs to talk or read about it?"

And so I will justify myself. First off, it is obvious that sexual behavior interests me a great deal personally. I have been a fantasist about sex all my life, and still am at a ripe old age. So it's not just "dirty old man" stuff. And insofar as my social and psychological imprint has allowed, I have taken part whenever I could.

Some of my readers may say at this point, "Well so what? Lots of people think about sex and some even participate. They don't necessarily yammer about it all the time."

That is also true, but my defense is that I yammer about all kinds of things, with pencil and paper, that is. I have this terrible compulsion to put things down in words. And when I consider a topic as important as sex, I just can't help myself. I don't write about everything. There are too many behaviors that are either too mundane, or otherwise uninteresting, that I have no compulsion to write about them at all. But sex is certainly not one such. It is the absolutely most vital behavior in the continuation of any species, including humans. The only other behavior that is more important, and that for the individual's continuity, is eating. And of course these two behaviors normally take precedence in reverse order. Animals do usually put more importance on eating food than on having sex, not excluding the human animal.

Even so, there are quite a few instances when animals, usually males, will literally waste themselves away by copulating to death.

And again to go back to male ribaldry, we used to talk about people "fucking their brains out." I'm afraid that is more fiction than reality. But to go to the elk which has copulated so much that it is then easily dispossessed by another male, or taken by a predator, it will have by then passed on much of its seed. The group will continue even if the individual will not. So if it is worth writing about technological, artistic, economic, or social behavior, it is certainly justified to write about copulation. It is very important, as well as interesting.

So why has there not been more put down through the years, at least in respectable circles? I can only speak of our own culture in this regard because though as an anthropologist I am always interested in other cultures, I don't honestly feel that I know very much about the sexual way of "others." And so what about ours, what about Western culture? For better or worse, in the last few centuries of Western dominance, sex has been restricted or taboo to talk or write about, at least outside scientific circles. A few "avant garde" writers have done so, but their products are still touchy in most social groups. Children have been taught to avoid saying much about it, much less doing anything. And adults have been reluctant to treat the subject openly. My father's sex talk was not unusual.

In English we evolved a double vocabulary for sexual things and occurrences, one Latin derived, the multi-syllabic words, the other from Anglo-Saxon, the four-letter words. There was a vulgar way of referring to sex organs and acts and a "nice" way. We were also taught to equate sex with uncleanliness. Sex was "dirty." There were "dirty movies," "dirty books," "dirty old men," and "just plain dirt," all referring to sex. There has been a tremendous avoidance of talking about sex as it really occurs. No wonder most of us grew up knowing so little about it. And even though there have been some changes in talking about sex, these certainly have not been earth-shaking. I suspect that most parents still avoid telling their children what really goes on. The primary source of information must still be the peer group. Our media is always in the business of keeping children from finding out what's going on. And of course we continue to hide the sex act. Even so, things have changed enough that something as down-to-earth and normal as barnyard watching can now be described in some detail, even if one is still trying to avoid four- letter words.

And so because of my personal attitude toward sex, as well as the change in attitude that has taken place during my lifetime in the social body, plus the significance of sex as a life process, I have long wanted to put down some of my ideas on the subject. Then by chance I raised the "horny hens" and I could resist no longer.

I must point out that the occurrence with the hens is a part of a larger change in my life. I got started as a barnyard watcher in my early years because of a lot of exposure to barnyards and domestic animals. Otherwise, I actually grew up as a city boy. However, I remembered those early years fondly, and especially my experience with animals. So after almost fifty years as an urbanite, I finally got a chance to go back to my roots. I bought a country place, Hilltop. And though it began as an avocado farm, I knew from the beginning that it would be for all kinds of animals and plants. And so it would be the greatest and probably last period of barnyard watching for me.

At Hilltop sex is everywhere, though it is generally seasonal. Living beings combine their genetic essences when the time is right for reproduction and growth, as all nature watchers know. The fact that humans have sex any time of the year is because through culture they have largely got independent of natural constraints. For other animals, in the northern hemisphere the time for sex is spring, because days are longer and there is more sunlight and warmth. And though barnyard watching is generally associated with animal life, plants of course need sex just as much in the exchange of pollen. There has been much less avoidance in recognizing and talking about plant sexuality, presumably because the exchange of sexual parts seems much more remote from the human act. There is nothing blatantly erotic in the sifting down of corn pollen to the silky tassels of the female part, or the transfer of pollen on the legs of bees. We do not see any liquidity or emotion or strenuous coupling movements, much less hear cries of ecstasy.

This is not to say that no peoples ever recognized the parallel between plant and animal sexuality. In fact, there is a fair amount of ethnographic information in anthropological literature about human use of such knowledge. Probably most often the use was magical. As I described in one of my other pieces, "Love Affair with Avocados," there is a kind of magic called sympathetic in which events performed ritually are believed to cause natural events to act in sympathy. Thus, water poured on the ground ritually will induce a rainstorm.

Sympathetic magical acts were described in considerable detail by Sir James Frazer, the 19th century guru of magic. And as I remember, a particular kind was a magical act to bring fertility to the fields in peasant society. A young man and woman would copulate in one of the fields and the surrounding plants would produce abundantly. In other societies growers used plant pollen or representations to promote abundance, cure illness, or produce other benefits. The Indians of the American Southwest were particularly well known for their use of corn pollen. Anyway, we who have a scientific outlook try not to use a magical approach. And lest one of my readers of "Love Affair" claims that I did, I must put down here that the love making was not being done with serious intent, for the magical aspects, that is. I had no real belief that our copulation would improve the productivity of the grove. And since I am now on this topic, I will add a new piece of information -- that the magic tree died while trees all around it flourished. Needless to say, I do not think the death of the tree decreased the productivity of the grove, as the act of copulation did not increase it. Rather I suspect the grove's productivity is due to fertilizer, water and pruning. I must say, however, that I did get a lot of fun out of the affair, mostly thinking and writing about it.

So, we scientific types really do take plant sex seriously. I watch all my fruit trees and do what can be done to promote a bountiful harvest. The actual pollination mostly takes care of itself with the help of insects or the wind. The most indefatigable workers in pollination are bees. Here in southern California they are out on warm days all year, carrying the pollen from flower to flower and from tree to tree. As with most workers, they are not selfless, doing this for the benefit of us growers. They are looking for nectar and in the process disseminate the sex components of the flowers. I have a hummingbird feeder outside my kitchen window which is frequently plagued by bees seeking sugar water, which to them is just as good as nectar.

Almost all bees are trucked to places near groves by commercial bee providers. These people get paid by owners of large groves. The bees are hauled around by truck in closed hives, deposited at designated places during the night, to begin work when the sun comes up. There has to be a lot of bee spillover, since I have never even been approached, nor has it been suggested to me that I make arrangements for a load of bees for myself. And they have always been numerous among my fruit trees and flowers. Other insects pollinate some plants. I have put in a lot of fruit trees from other parts of the world and almost all have been accepted by the bees or other local insects. One, however, has not, and has to be taken care of by hand. This is the cherimoya or custard apple from South America. It produces a delicious fruit, pollinated by a small beetle in its home territory. However, it has not been taken over by our pollination work-horses and produces poorly here unless man steps in. In the past, one of the innumerable plant experimenters in California discovered that the trick was to hand pollinate the flowers. These are inconspicuous little green jobs, which may be the problem. They lack the splashiness of most other flowers. Anyway, the bees don't come to them. So, as I learned early, in season the grower has to collect the pollen in the afternoon with an artist's paint brush, store it in a small container, then dab it on the female part in the morning. The tree will set much more fruit with this help.

Jessica thought many of my activities in the country were odd, if not bizarre, when we first met. In our courting days she developed the habit of calling me in the morning when she was in the city and I was at Hilltop. One of her usual openings when I picked up the phone was to ask me what I was doing.

I suspect when I got used to this question, I developed a pattern of laying it on. I had learned by this time that she, like most of the ladies who were attracted to me, liked men who were unusual in some way. They, like most humans, were perpetually trying to avoid boredom. I had learned to accommodate them. This was easy enough if I drew on my chosen subject, anthropology. There was enough bizarre information in the ethnographies of the world to keep legions of ladies beguiled. I drew on them for this purpose. However, I had learned that activities at Hilltop would also serve. Just living on a farm part-time generated a lot of new ideas. And being a long time barnyard watcher, the possibilities were even greater. Anyway, I would often regale Jessica with oddments when she made her morning call. I learned that not only was she impressed by much that I told her, but that if it were bizarre enough, she would spread it far and wide, to her colleagues at the hospital unit where she worked, and to her several phone friends, some of whom she talked to every day. So one day when I hurried into the house to get her call, just coming from pollinating the tree, and she posed her usual question, I said, "I was fucking the cherimoya tree."

I used the Anglo-Saxon word specifically because I knew Jessica was no prude and also she was easily titillated by a sexual topic. Neither of us customarily used four letter words in our normal conversations. However, I thought the shock value was worth the choice this time. Furthermore, I thought she would really be interested. And finally I just could not imagine using a Latin derived circumlocution. Granted I could have said "pollinated," which is Latin derived, and the most commonly used term among growers. But this choice would have emphasized the separation of plants from animals which I did not want to do. Like Darwin, I wanted to emphasize the common element of sex shared by both. I could have used the Latin derived term, copulation, but that was too stilted for our conversation. So, I opted for the most blatant word for sexual intercourse, though admittedly not usually applied to plants.

I must confess also that I get a kick from using four-letter words. They have of course been acceptable in masculine conversation all my life and in the last twenty years have been increasingly used in bisexual talk. Anyway I knew that they did not bother Jessica. Still, she did not understand me exactly in regard to the cherimoya. She said, "What do you mean?"

I could visualize her imagining me trying to penetrate a knot-hole. I vaguely remember boyhood jokes about men trying to screw postholes. Anyway, I didn't follow that up, instead saying, "Oh, I'm not down there trying to have intercourse with a tree. I am pollinating it."

Then I explained the treatment of the cherimoya, which obviously intrigued her. Not too long after, I learned that she had spread the story widely among her colleagues and friends. I wondered if she had used the word, "fuck" in the telling with any. I knew she could do that. Incidentally, she became very fond of eating the cherimoya fruit and has talked it up with many people . Anyway, although all the plants at Hilltop have their sexual component, none makes a better story than the cherimoya.

As a husbandman, I try to keep tabs on the flowering of the

fruit trees and garden plants. Sometimes the fruit trees, and particularly the avocados, are stressed at the time of pollination, usually from a heat wave or cold snap. I can do little about such. Using the wrong fertilizer on the garden may cause plants, particularly tomatoes, to grow leaves instead of fruit. That can he corrected by changing fertilizer.

But so much for plants, because their pollination of one another is hardly titillating. The true basis of barnyard watching is animals, because what they do is much more like humans. And so what about the animals at Hilltop? I have already discussed the chickens which have adopted me as a stand-in for a rooster, because as Jessica said when I told her about it, "To the chickens you're the only action in town."

We had some ducks a couple of years ago, one of which was a drake. When they got big, they went at it in duck fashion. They had a plastic child's pool full of water. And when the mood struck, which seemed often, they swam around and around in courting excitement until he got in position on top of his mate, whereupon he literally tread her under the water in a mounting frenzy. He would keep at it long enough that she would be quite out of breath when she popped up. Elsewhere I have seen several drakes mounting one female in this fashion and seeming to keep her under water for an impossible interval. Duck copulation can hardly be seen as anything but rape in human terms, though the female certainly participates in the preliminaries.

I have an indoor aviary, populated with several kinds of birds, but mostly doves. They are very active sexually, the male going through an elaborate tail-spread prance until the female squats to make a platform like the chickens. And in parallel fashion, he jumps up and inserts as fast as he can before falling off. After which they both puff up and dance a little in self-congratulation. There is no suggestion of rape, both parties taking part fully. Eggs appeared from these unions, but after raising two or three pair, I got in the habit of tossing new ones out before chicks were hatched. Doves are not highly valued birds and it was a problem getting rid of the young ones.

There are many wild birds around Hilltop which have their own way, which for most is not particularly spectacular. All one can say is "It must work."

Many are usually out of sight, not for modesty as with humans, but simply because they can fly to inconspicuous places easily and for safety do not want to be too close to people. I have seen pairs of hawks going through wild gyrations in the sky, which I presumed was courtship display.

There are quite a few reptiles around the place, mostly lizards, and I see them during the warm season going through a kind of dance behavior which I believe is courtship. Probably the most spectacular courtship-coupling behavior I have witnessed was of gopher snakes and which I have already mentioned. Gopher snakes are about five feet long and completely harmless to humans. Obviously this is not so for gophers and I presume other rodents. I see them around throughout the warm period, usually lying motionless or moving slowly across the ground. They are easy to catch and do not bite so I have picked up quite a few by tail, and after showing them to Jessica or someone else who was there at the time, released them.

One day in late spring, I went into the garage and was surprised to see a gopher snake's head pop out through the grill on the front of the pick-up. It was quickly followed by another. As I watched in amazement, up to half a dozen slithered in and out, wrapped around one another or separated, disappearing and reappearing. It was a veritable orgy. I have never seen these or other snakes act this way in the wild, though I have seen TV sequences of a kind of garter snake which goes into a sexual frenzy. The snakes I see usually do nothing more active than slowly crawl away.

We have fish and turtles in the house but they give no sexual performances. However, the fish in the outside pond, the koi, or decorative carp, put on quite a show for a couple of weeks in spring. When the water gets warm enough, the females, which are noticeably blockier than the males, give off signals that they are ready to mate. Whereupon males get behind them in groups of up to half a dozen and go careening around the pond, the males jostling the females to make them lay so they can spread their sperm over the eggs.

Koi breeders tell me that left alone, the adult fish will eat the eggs. The usual practice is to put breeding adults into small separate ponds for eight hours or so and then whisk them out before they devour their own eggs. Some water plants are said to be useful to protect the eggs and baby fish. I do have a good growth of water hyacinths and the result is that I see small colored fish some weeks after the breeding frenzy, but none seem to survive longer. Evidently the hyacinths do not provide enough protection. But since I am not in the business of breeding koi, this outcome is no particular problem. In fact, I have awakened some nights with a start, thinking that all the koi eggs were hatching and the fish were growing up. The pond would become a seething mass of baby fish, which I would go crazy trying to take care of. Fortunately, this does not happen.

And finally I come to the dogs, the two fox terriers and the Doberman. Pete and Repete, the Jack Russell terriers, are long established at Hilltop and are physically different from one another, despite being litter brothers. They vary considerably in build and temperament, sexual as well as other. Darwin would have doted on them. Pete is slightly bigger and early on established dominance over smaller Repete. Pete's usual procedure when they were growing up was to mount Repete occasionally, to show his dominance rather than to copulate. As often as not, he mounted Repete at the front or side. But they were brothers and got along well enough.

Of the two, Pete was the better endowed. He was also more active sexually. Hilltop is fairly isolated and not over-run with dogs. So, presumably Pete got few chances. And so far as I know, he kept his legs crossed, so to speak, for a long time. But his testosterone finally went up, or some other thing happened which got him stirred up. When he was three or four years old he disappeared one day. By this time Jessica was coming to Hilltop regularly, and she got quite concerned. So after Pete was absent a few days, we put signs on telephone poles about our lost dog. It took only a few days before we got a call from a neighbor across the canyon, two-three miles away. When we got there, we quickly spotted Pete settled down in a seraglio of small female dogs, in a sexy dog's heaven, it would seem.

But we asserted our human rights and brought him home, to be taken to the vet for castration the next day. In this case, a dog's life was not so hot. Pete lost his interest in lady dogs as his little scrotum withered. He also almost doubled in weight. Like others before him, human and non-human, he made up his loss of sex by increasing his food intake. He became round and comfortable with no further interest in lady dogs. He also lost some of his aggressiveness.

Repete had no such problem. He was always slighter than Pete and his family jewels were always diminutive. In fact, his scrotum was about the same as Pete's after the castration. Anyway, Repete never had any drive to get with lady dogs, and never trotted off in their search.

Then there was Rhett, the Dobie. He was named after Rhett Butler by Jessica who generally named animals after showbiz people. He was a big handsome fellow, quite muscular and quite well endowed. However, he was almost totally oriented toward people and very amiable. He made the best testimony that Dobermans have been given a very bad press. I could not imagine him attacking anyone. The rat terriers had. And in fitting with his people orientation, I know of only one instance when he showed interest in a female dog. And that was very brief.

And that's the way it was at Hilltop. The animals did their thing, as people do, keeping their kind going and perchance having some pleasure in the process. And I watched my version of a barnyard. So, while I think of it, perhaps I will step outside and give the horny hens a few flat-hand pats. I am sure they would enjoy it.

19

Love Magic

    Frequently an anthropologist will be told that the subject under discussion would undoubtedly be interesting to his kind. Anthropogs (a shortening of the term that I developed during my teaching days) are thought to be concerned with the bizarre. This is partly true, I'm sure, if for no other reason than that anthropology is cross-cultural. The normal anthropog looks at other peoples' ways and thus encounters many alternatives for dealing with human problems. A lot of these variations on the cultural theme seem odd to those who have not ventured far from their primary cultural shells. And frequently what the well brain-washed find odd in human experience will be considered by the anthropog to be simply an alternative for dealing with a human need. As a matter of fact, probably the main thrust of most anthropology is to spread the message of how many possibilities there are for dealing with human problems.

    And yet this cross-cultural outlook will not be accepted easily by most outside the field since the very nature of cultural brain-washing is to imprint on the receiver's mind the message that his is the right way. The well brain-washed will not question what he has been taught in the early part of his life. Thus, when he learns of alternatives, he will get rid of that bothersome fact by putting the other way into the category of bizarre behavior. This procedure is the fundamental basis of ethnocentrism, that is, judging customs of others according to the standards of one's own culture.

It isn't so hard to accept alternatives if they do not differ markedly from one's own ways. But the greater the difference, the harder it is to accept the other's ways. Thus, the standard American has little difficulty in accepting the idea that people of the Middle East eat goat meat. After all, the American eats mutton or sheep meat. But it is not so easy for the American meat eater to accept the idea of eating dog meat, since he doesn't eat the meat of any animal like it. And it is probably even harder for him to accept the alternative of eating fat grubs or grasshoppers, both nutritious animal proteins, but totally outside the American's experience. Also, though he may criticize the custom in his own culture of bombing people into devastation, the normal Europerson accepts this as a necessary evil. Ritual killing, on the other hand, is soundly condemned by the Europerson. The Spanish conquistadors were highly indignant and stamped out the human sacrifice of the Aztecs as soon as they got in control, which of course they did through the standard patterns of Euro-war of the time. And at the same time the Spanish were burning large numbers of people at the stake in the Inquisition. They considered their form of killing as moral while they condemned the Aztec form as barbaric. This attitude has hardly weakened. Great indignation was waxed in 1989 by Americans when the ritual killing of about fourteen persons was discovered on the Texas-Mexican border at the same time that official Americans were sending conventional arms to several guerilla armies in standard warfare against established authorities identified as non-democratic. Like other behaviors, the kind of killing that humans approve is relative to the particular culture.

Anthropogs, and some other social scientists, get their information mostly in two ways. They either watch what is going on or they ask questions and then listen, frequently a combination of both. Looking is the basic way of getting information through the scientific method generally. Listening is done less frequently, because man is a seeing animal more than a hearing animal. All the sciences concerned with non-human things depend almost exclusively on looking. If one wants to study something about the universe, one has to look at bodies in space through a telescope, or less often to listen to sounds bounced off celestial objects. Yes, a brilliant theoretician, like Einstein, could work out an elaborate theory of space, time and energy simply through noodling, but the ultimate proof of the theory still depended on seeing through a telescope the relative position of some stars. Also to know about the smallest things, even those not visible to the unaided eye, one has to look, with special instruments of course. And if one wants to know something about the arrangements of the human body, one has to look at its carved- up remains (autopsy). Even learning about animal behavior now largely depends on observation. Jane Goodall learned that male chimpanzees were basically aggressive animals only by watching them for twenty years.

But from the cultural animal there is another way of getting information, namely listening to him/her. He talks about what he does. Now, some reader is bound to say that other creatures communicate, make sounds to transmit messages. And it is so. The other animals certainly do communicate. But without going into the details, it is fair to point out that language, symbolic speech, is a very different way of communicating from that of any other animal. In short, it gives very precise information, as well as having a lot of other special qualities. And language permits storage and regurgitation of events. It enables the user to play with time. It not only can extend itself into the future, describing events that have not yet happened (in religious or scientific predictions), but it can also describe what has already happened, both imaginary or real. Language is a time extender. Is it any wonder that anthropogs and other social scientists early became hooked on using language for collecting information? In short, they learned to ask people what had happened.

Using language for getting information is a two-edged sword, however. The good side is that one can get an idea quickly of events that have already taken place. The answer to "How many hours of TV do you watch each week?" is much easier come by than sitting around all week watching the person, a stop-watch in hand. Certainly the information is less precise, since the person answering may have a poor memory. But even so, the advantage of getting a large number of responses in the time it would take to watch seems worth the trade-off to most who study people. There are other problems, however. There are human experiences which are difficult to watch. One such is sex, another is crime. Few humans want social scientists around watching when they are having sexual intercourse or robbing banks. Even these taboos have been breached, however, though most information we get on these and other taboo subjects comes from interviews. When I agreed to give information to a Kinsey interviewer on my sex life, I do not think I would have agreed to let him observe me in action. So typical of most sex studies, the Kinsey reports were done with very little observation.

Most such interview information is got through question and answer sessions, normally set up by the social scientist. He asks the questions, they give the answers, and hopefully with a fair degree of truthfulness. However, after many such interviews, the long-time social scientist gets to the point that he cannot help collecting info, even when he's not in a regular interview session. Anything that occurs in an ordinary conversation becomes grist for his mill.

The moral may be, "Be cautious what you say in the presence of a social scientist if you don't want to end up as a statistic or example."

I remember many such conversational bits from a field study I once did in Trinidad. I was particularly interested in religious and magical beliefs and practices which the locals seemed to be very concerned with. When they learned of my interest, several young men from the local town would come to my house night after night to regale me with stories about "jumbies," punctuating their tales with hearty swallows of puncheon rum, the local firewater. I hardly had to do more than listen, then afterwards write down all I could remember. However, reference to spirits could come out in any conversation. My wife and I had a school teacher couple as friends with whom we traveled about on auto trips. Both were devout Christians and denied that there existed such things as "jumbies" or any other kind of non-Christian spirit. The two of them were sitting in the back seat of our little Vauxhall once while Jacob, the husband, was so debunking. I kept mostly quiet, saying only enough to keep him going. I had long learned that the good information collector is a listener, not a talker. Anyway, Jacob was a little scornful, saying, "You know these local Indians (East, that is) believe all kinds of strange things. Any little unusual thing happens and they'll say it was caused by a "jumbie."

His wife, Isabel, agreed. "Oh yes, they are so superstitious."

I kept my counsel and after a pause, Jacob said, "Of course one does see some strange things."

My ears perked up as I waited. "Like the time we saw the headless people in the back seat of the car."

With all the cool I could muster, I asked, "Headless? How's that?

Jacob, "Well, we were driving home late one night when we heard this whispering in the back seat. And when we finally looked around, we saw these two people, wrapped in gauzy cloth, but headless."

I of course wanted to ask how the creatures without heads, and thus mouths, could whisper to one another but I had learned by then that supernatural events in all cultures were usually not to be explained rationally. After all, even in Christianity how can one rationally accept the idea that a person could hold back an ocean with a stick or that another could be swallowed by a whale and come out unharmed. By insisting on rational explanations of such events, the ethnographer would frequently turn off the source of information. So I simply asked Jacob, "What did you do?"

He answered very seriously. "Well, I can tell you we were really concerned with those headless people. But I finally stopped the car, got out and opened the back door, and they went away. Just vanished."

Oh no, Jacob and Isabel did not believe in spirits! I was a little proud of myself that I had not interrupted his account to insist on a rational explanation. But that's the way we dedicated ethnographers finally became. We listened to all kinds of seemingly outlandish tales when we were formally interviewing but also in ordinary conversations. We could hardly help ourselves. That was our own kind of brainwashing.

Forty years later Jessica told me this tale, which she had learned from one of her colleagues in the hospital. This nurse had a friend who had been talking to Jessica about the ins and outs of marital fidelity. And this friend, who I shall call Betty, told Jessica's colleague, "But I have no problem with my husband. He will never go out on me".

Whereupon the colleague, who I will call Helen, said, "But how can you be so sure?"

Whereupon Betty said, "I'll tell you how. I fix his coffee ."
Whereupon Helen, "And how do you do that?'
"I line his coffee pot with a piece of menstrual pad." Pause. "Used, that is."

Helen gasped, "Oh my gosh." Another pause. "But even if you do, you ought to be careful. What if he heard you?"

Jessica, "And that's what Helen said happened. He was listening."

I couldn't restrain myself any longer so I said, "And then what did she say happened?"

Jessica's eyes were sparkling. She delighted in stories about love, even bizarre ones. She, "Helen said that when Betty's husband overheard the bit about the menstrual pad, he went off his rocker. He went out and got some acid which he threw into her face."

I, "Didn't something happen to him? Wasn't he prosecuted?"

Jessica, "Oh yes, he was tried and sentenced to prison."

I found one part of the story particularly intriguing. What was Betty's cultural background? After all, the idea of menstrual blood for love magic was widespread throughout the world. It had been included in the category of contagious magic and described in considerable detail by Sir James Frazer. According to this idea, the essence of a person remains in his/her parts, even when separated from the body. Thus, the intimate substances of the body which come out/off, fingernails, hair, spittle, feces, and menstrual blood can be used magically, to make the person do what the magician wants. A wife could thus control her errant husband. I said to Jessica, "Did you know what Betty's ethnic background was? Could she have lived in or known about another culture?"

Jessica, "Not that I know of."

Another thing intrigued me, the husband's reaction. And though I know that in some other cultures where magic was practiced, that victims or their relatives sometimes resorted to violence for revenge, more often they would use counter-magic. Only in a culture where people would see the act as pure viciousness, with no idea of it's magical aspects, would they take such drastic action as to throw acid on the guilty person. Betty's husband probably thought the use of the menstrual pad totally depraved.

Being a dedicated cross-culturist, I couldn't help comparing the occurrence in the heart of civilization, Los Angeles, to what I knew was done elsewhere. So back to Trinidad, land of jumbies, magic and all kinds of supernatural goings-on. The Trinidadians were deep into love magic, as I learned early in my stay there. The boys from Debe, the ones who regaled me with tales of jumbies, also early described occurrences of magic. A very common type was love magic and a widely used procedure was called "tying someone to the bed," the errant spouse, that is. It turned out that a lot of married partners were messing around, particularly the men. But to keep the situation from getting out of hand, the wife would fix a magical preparation which included some intimate substance from her body. This would be buried in the yard in a small bottle. The wayward husband could not pass over the bottle without getting zapped magically. As best I remember, this meant he would get dizzy or sick or otherwise feel funny enough that he wouldn't want to continue on to visit his girl friend.

At that time I was not as aware of magical practices as I became later and all these stories sounded a little preposterous to me, even for Trinidad. Anyway, I followed through, asking if the husband took all this lying down, so to speak. Didn't he have any counter-measures if he suspected his wife was trying to exert her will over him magically?

"Oh sure," the Debe boys assured me. "He could use various kinds of counter-magic. And he even had a sure-fire way of finding out if she was doing it.

"And what was that?" I asked after taking a small belt of the puncheon rum then being passed to me.

"Well, there is a special kind of chicken, one with its feathers growing in a curl. In Trinidad it is called a frizzle-fowl because it looks like it was passed through a hot fire where all its feathers got curled. A man could keep one such in his yard because it had the ability of locating the magic. The bird would station itself over the location, its beak pointing at the spot where the object was buried."

I was so inundated by tales of magical and spiritual goings-on by this time that I just couldn't maintain my credulity. I said so, trying not to offend. Even so, such a procedure is hardly recommended for the inquisitive ethnographer. Anyway, I quickly collected my cool again and asked the Debe boys to show me these things when they could, the chicken called a frizzle-fowl and the magic ingredients. One fellow responded quickly, "No problem, because I'm almost sure my wife is using magic against me right now." He grinned conspiratorially, "'I do have a girlfriend."

And so it happened. In a few days I was shown the chicken which looked as bizarre as it had been described, and which I duly photographed. Not long after, the errant husband showed up at my house, carrying an aspirin bottle he had just dug up. At my request he opened it carefully and laid the contents on a white cloth which I provided. Included was a folded piece of paper rolled up, the letters INRI written horizontally and vertically, a few seeds, and a small piece of discolored cloth. I photographed the contents.

How different the reaction of the Trinidadian to that of the magicked husband in Los Angeles. In Trinidad, where magic was endemic, one countered like with like. There was no need to resort to violence.

And so ethnographers learn to listen, in formal interviews, but also anytime a tale is told.

20

The Real Life Church

I was started out in traditional fashion in the Catholic Church. My earliest memories include getting cleaned up on Sunday and marching off to church with my parents and/or sisters. When I was nine or ten years old I began delivering newspapers on Sunday and was then expected to go alone to early Mass before doing my paper route. It was painful to get ready for the five o'clock Mass. Otherwise, I found the service mostly boring. But as I got older, I began taking exception to points in the sermons, primarily on the grounds of credibility. As I learned more directly of the small world I was living in, and more of it indirectly from the books I was reading, I found it progressively more difficult to accept so many bizarre occurrences as having really taken place: a man who was swallowed by a whale and survived, another who parted an ocean by sleight of hand, and still another who called down showers of snakes, toads and frogs. These things just didn't jive with the world I was seeing and reading about. And I am afraid the explanation that they were miracles and had to be accepted on faith was harder and harder to accept.

I wasn't the only doubter. I remember my rough-hewn uncles at their farm church in southern Indiana, arguing with one another after Sunday Mass as to whether such and such a "miracle" could possibly have taken place. And these were tillers of the soil who did not even read books, which is probably why they stayed with the church anyway.

My case was different. I was raised in the city in a neighborhood where Catholics were laughed at, and who read books, many outside the church-approved list. Although I did not yet know the word, I had already learned to look at the world empirically. Anyway, by the time I got to college I was not much of a believer. And as I picked up more naturalistic science, especially from anthropology, what little belief I had rapidly dissipated.

So without really looking for a new faith, I was open to other religions which I learned about in reading or on field trips, mainly to Asia. I never found Judaism or Islam very interesting, probably because they and Christianity are all variations on the same theme. But Hinduism and Buddhism, variations on another theme, were more interesting, mainly because they incorporated all forms of life into their creeds and worship. They did not set man up on his own special pedestal as had been done in the JCI Triad (Judaism-Christianity-Islam).

And what has this to do with the topic at hand? Believe it or not, it is an introduction to one man's religion which I call a Real Life Church. This isn't to say that after rejection of my traditional faith, and exposure to all the kinds that are described in the ethnographies of the world, that I fell back onto a dependence on a belief in the supernatural. No, I am honestly as agnostic as I was when I walked away from the Catholic Church. I never really found another belief system that was convincing enough to adopt, though I must admit I didn't look too hard. In any event, I honestly do not know where I and all the rest of the life forms came from. I do not know how they were created, nor what their destination is, much less any kind of supernatural master plan. Darwinism is enough for me, and if there is a primal cause, so be it. And if there is no continued consciousness beyond death, which I suspect is the case, so be that also. Despite the declarations of the sermonizers of my youth, I find no discomfort in this lack of belief in a traditional system.

So why then a personal church? The Real Life Church is actually a metaphor which incorporates several aspects of religiosity but without the supernaturalism. It is a church with no concerned god, as is most appropriate for an agnostic. In this sense it is more like Buddhism than any other major religion, and I have thought that an agnostic would have little difficulty in accepting that creed.

The only force in which I can find meaning is life itself. Since most life as we know it is dependent on sun power, I think it would be reasonable to be a sun worshipper. This has been done in many religions. I find the idea reasonable except for one caveat. I know of no reason why life had to emerge. There was nothing inevitable which compelled those primitive molecules to have developed the capability to reproduce themselves. And if they had not, the great fiery orb would have continued sending its life-supporting rays onto an inert planet forever. The dead earth would have continued rotating until the sun went out or blew up, throughout lacking the variety of living forms that evolved from those single- celled creatures, our ancestors. And thus I find life itself as the force beyond all others that a non-believer in traditional faiths could believe in.

The natives in the South Pacific had supernatural systems based on a belief in widespread generalized power, called "mana." Also it has been discovered by anthropologists to have existed among many other peoples. I think of it as a kind of supernatural electricity. It could be used to turn on the lights of the world or to kill people. "Mana" is a little like the life force I am talking about, though it includes non-living as well as living beings. But still there are many parallels.

And so what is the Real Life Church, particularly in its individual aspect? I see its prime force as the reproductive capacity of living things, so far known only on our planet. It has existed for three billion years or so, when inanimate became animate. The idea assumes that if there is a god, he is high and remote. He got the system started and then retreated into the background, as described in the cultures of many pre-urban peoples. The prophets are Charles Darwin, Alfred Wallace and the later evolutionists. There is no clergy. There is no Mecca or Jerusalem or Benares or the San Francisco Mountains as a focus of faith. And there are no structures to house the faithful in rites of intensification (the mass, "namaaz," Buddha's birthday, Xmas). Instead the entire world of the living serves as the altar and a service takes place each time life succeeds. A religious occurrence, a miracle if you will, takes place each time a rabbit escapes a coyote. It also occurs each time a coyote catches a rabbit. In the first instance, rabbitry is enhanced, in the second coyotery, both of course being legitimate life forms. If there are priests in the Real Life Church, they are naturalists or men from other scientific disciplines who study and try to explain how life works. The holy book is <u>The Origin of Species</u>. The devotees are those who worship at their own particular altar of life. I am one and my altar is Hilltop.

What are the acts of worship in the Real Life Church? In the first place, the devotee must do all he can to survive as an individual, not from egocentrism but because without life all other acts of worship are impossible. His will to live must be unquenchable even though he knows the struggle will ultimately be lost. Those with major disabilities, but who still struggle to live, are the most obvious worshipers at the altar of life.

When I was in graduate school, a fellow student was a quadriplegic. He could only get about by being pushed in a wheelchair, and he could barely be understood, his vocal system was so badly damaged. He could not hold his head upright but still managed to write notes, his head sideways on the wheelchair support, his mouth drooling. He participated in class question and answer sessions, though he usually had to repeat questions several times while the professor and other students would wait. Still, he persevered until he got his Ph.D. in anthropology. He then got married to a girl who was also disabled. He made himself a specialist in the literature of Russian peasantry, I'm sure learning Russian. He wrote many articles and books on Russian peasant life, and got a faculty position in a university.

Then I had a colleague who after getting polio in his twenties, and becoming paralyzed from the hips down, also went on to get his Ph.D. in anthropology. He launched an academic career, did extensive field work in a wheelchair, got married to a physically normal lady, sired and raised a family of several children, became chairman of his university department and wrote much, almost all from a wheelchair.

And I must cite one of my long-time literary favorites, Prescott, the author of <u>The Conquest of Mexico</u> and <u>The Conquest of</u>

Peru. He was a man who lived in the 19th century and who was almost blind. However, he was determined to do something productive in his lifetime. Fortunately he had enough inherited wealth that he did not need to work for pay. Moreover, he had enough to hire a permanent reader. He chose to have read to him all the works in the Spanish archives on the two great New World empires, including their conquest by Cortez and Pizarro. Then he dictated an ethnographic description of the peoples as good as could be known from the written works of the time, plus a stirring account of the conquests by those soldiers of fortune. After translation back into Spanish, the accounts of Prescott have continued to be widely read in the countries where the events took place, as well as in English. Prescott never went to Mexico or Peru. Nevertheless, all the physical descriptions seem to be very accurate, even though they came from a man who could not see the world around him.

Most persons have done nothing so impressive. Most have not had to. And yet almost all who worship at the altar of life have participated in miracles of consequence. My own took place later in life, as is not unusual of course. It was a stroke, which I got when I was 59. And even while life-saving procedures were being administered to me in the emergency room, I was already absolutely intent on recovering, which I did almost completely. (See "The Stroke")

The true devotee fosters any form of life, not simply his own. Thus all creatures are supported. The major religions of the world, as well as most tribal ones, can be logically divided into "life affirming" or "life using" systems. The JCI Triad (Judaism-Christianity-Islam) are "life using" systems. In their creeds all other forms of life exist for the benefit of mankind and he can use them as he wishes, in everything from eating them on a mass scale, sacrificing them to his gods, to using them for experimentation. To facilitate this differential treatment, the JCI Triad divests other creatures of a soul, a fictional entity that is supposed to exist in humans. On the other hand, the JHI Triad (Jainism, Hinduism, Buddhism) treats all living forms as similar, even if not equal, in their supernatural characteristics. All creatures have souls and the living individual is nothing more than the current manifestation. One moves up and down the ladder of life through death and rebirth but one does not live a separate existence from non-human creatures. The consequences are many but an important one is that believers of the JHI Triad are less likely to heedlessly destroy other creatures. Thus, vegetarianism is much more widespread and many kinds of non-human creatures live side by side with the human ones. This is especially true in India because of the influence of both Jainism and Hinduism. Buddhism, while still having its life affirming aspects when it was carried to Eastern Asia, did undergo many changes in the direction of "using" in its new homeland.

Probably most tribal cultures are of the life affirming variety, or at least life incorporating. Most of them lacked the technological capability to bring about major destruction to the life forms in their environment. And since they lived more intimately with other creatures, and were more dependent on them, they tended to incorporate them into their religious systems. Popularly we call this nature worship, in anthropology it is animism. So even when animists killed a bear, salmon or seal to eat, they more often than not gave it a ritual send-off. All creatures had something corresponding to a soul which the tribal would try to placate ritually, even when he ate the body.

There have been many consequences from the differences of these systems. The life using systems have tended to be more destructive of life and the environment. There has been nothing in the supernatural belief systems of the JCI Triad to prevent massive environmental destruction, and probably the greatest ecological disasters have occurred in cultures influenced by these systems. Of course the sheer technological capacity of Euroman after industrialization enabled him to do much more destruction than was possible for non-industrial cultures. But also there was nothing in the religious system that inhibited such destruction. The land, and everything on it, had been created for man's exclusive use. Much of the destruction was thoughtless. Since he was an environment user, it was logical to wipe out species because they seemed to be in the way or because it was easy. And so the dodo, passenger pigeon, Carolina parakeet, buffalo, grizzly bear, and many more were driven near or to extinction.

Despite the new ecological philosophy, the attitude of destroying other creatures if they can't be used directly still remains. Most animals are popularly thought of as belonging to one of four categories: economically useful, domestic animals that are eaten or otherwise used; game, wild animals that once had economic usefulness but now satisfy recreational needs; pets, animals with which one establishes emotional bonds; and pests, animals from which we get no direct benefit and which at the least irritate us. We rarely think of animals in a neutral category.

When I was young, my uncles used to run in the house for a shotgun each time they saw a hawk flying over. At that time, farmers would drape on their barbed wire fences the carcasses, wings outspread, of hawks they had shot. My uncles, and I'm sure many others, called all hawks "chicken hawks." When they got tractors, my uncles would carry shotguns along when they went out to do field work so they could shoot any foxes they might see. The larger predators had long been killed off. Now it is generally thought by ecologists that these predatory animals are beneficial to the whole wild environment. They usually live off rodents and other animals that are destructive to crops. I suspect that the most common reaction to insects in our society still is to squash them. Spiders are rarely let be, even though they are highly beneficial predators of the insect world. Ants too make significant contributions to life, especially in pollination and seed distribution. It still comes as a great surprise to Westerners to learn that devout Jains, the most assiduous life affirmers, would feed ants with sugar. When I put a jar of sugar or can with some honey droplets remaining for the ants to finish, Jessica found it droll indeed.

This is not to say that I protect ants or spiders as much as a Jain might. I am too Western for that. Even if I do admire the life affirming characteristics of Jains, I am not a believer, and it is unlikely that I ever will be. I make choices and more often than not for my benefit. When Jessica would be bothered by a spider in the house, it was no problem to pick up the creature and put it outside. However, when hummingbirds and ants come into conflict, I choose hummingbirds. I try to feed all kinds of creatures at Hilltop and one is the hummingbird. I keep a container of sugar water outside the kitchen window. It is hanging on a wire for the convenience of the birds but also to keep it away from ants. Even so, they periodically find the jar and come down the wire. Ants have no problem driving hummingbirds away. In order to prevent this, I spray insecticide on the wire. I have also used Diazinon to spray the ground around plants or tree trunks where the fruit was being attacked by ants.

I don't have any objection to the use of chemicals, whether for maintaining human life or for controlling insects. After all, chemistry is a part of the empirical thought base of modern mankind. If we take this part of his technology away, humankind will be greatly diminished. I do however think that the selective use of chemicals and other technology is the way to go. Broadcast procedures with little knowledge of the consequences have frequently been harmful. But what I, as a devotee of the Real Life Church, object to more than anything is the thoughtless destruction of life. Thus, I am hardly in favor of shooting "chicken hawks," foxes, or coyotes, or whacking snakes to death, or crushing insects simply because they are there. I will admit that I did these things in my callous youth. But that was before I became a devotee of the Church.

I witnessed an occurrence at Hilltop a few years ago which really pissed me off. George, who was then a roomer at Hilltop, was outside talking to me one sunny day and began to methodically step on ants that were milling around on the driveway under where we were standing. He would put his foot on one and methodically press and turn the shoe sideways to grind up the pinned ant. I am afraid I became so distracted I couldn't keep up the conversation and I finally said, "Why are you doing that George? They're not hurting anything."

Whereupon, "They're just ants."

He was killing them only because they were insects and he could. How thoughtless of life the Christian heritage has made most of us, even those that do not go to church. Of course, the whole history of environmental destruction on our planet is part and parcel of this same heedless or ignorant kind of killing. Anyway, as devotees of the Real Life Church, we try to maintain life in all its manifestations, from the lowliest insect to the grandest predator, from ants to coyotes. I heard that there was a cougar in our area, and I thrilled at how wonderful it would be to see one of the big cats sloping up the hill in all its grace. Unfortunately, the grandest wild cat we have had for sure hereabouts was a bobcat. He wiped out my chicken flock, and though I was sad for that, I would not conceive of doing anything harmful to the bobtail.

Not too long before, I was visiting a neighbor near Fallbrook who enjoys raising animals, though without being a devotee of the Church. He is an old fellow, even more than me, and a retired factory owner from the Los Angeles area. He kept a wide variety of animals, including buffalo, emu, many kinds of fowl, and koi. I met him because of the latter. It was his greatest delight to raise the young of these various creatures, which he would then sell. To him animals existed to be used. When I was there that time a breeder of exotic birds was also there. They got to discussing the trap the man had in his pickup and how they had just caught a bobcat.

My ears perked up and I said, "So what did you do, take it up to the mountains and release it?" With a grin, Fred, my neighbor, said, "Oh no, he was taken to the taxidermist."

I became angry very quickly though I tried to contain it. "You mean you killed it?"

"Yeah, I shot it with a .22"

I had ravaged the countryside with a .22 in my youth, for which I have been doing penance ever since joining the Fellowship, so I said, "What did you do that for? Had it killed anything?"

"No, but they can, all kinds of birds and small animals."

I'm afraid I flipped. "My god, I can't believe it. Here we are living in a country where we've killed off all the big animals. And then we kill one of the few left of a species that has managed to survive. Jesus, I can't believe it."

Fred quickly tried to justify himself. "It probably had killed some local fowl or other animals. And it's legal for a farmer to kill marauding animals." I was almost blind with anger, visualizing that handsome cat turned into a taxidermist's mannikin. To solve my dilemma, I quickly got off the topic and made an excuse to leave. Needless to say, Fred is no devotee of the Church. Though he likes to keep and breed animals, he is strictly a user. If he can't get some practical benefit from an animal, he prefers to kill it. And with that belief, he and most of our ancestors denuded the land of wild creatures.

At Hilltop we welcome wild creatures, even if they are predators. There is no way that we would kill one of the many hawks hereabouts, even though some would pick up a chicken if they could. I have seen them eyeing the koi hungrily while perched on the bridge railing. There are no guns at Hilltop. The only traps are a couple for gophers when they occasionally get to wiping out rows of garden plants. One recently destroyed 6-8 cauliflower plants and still escaped untouched. Gophers elsewhere are free to do what they want, and they are all about.

I feed wild birds, sugar water to hummers, seeds to doves, finches and other seed eaters. Inadvertently I feed avocados to a number of animals, especially tree rats, ground squirrels, and coyotes. They simply come and help themselves. They do not eat many. A fair number are cut loose by squirrels or rats who gnaw through the stems, whereupon the fruit drops. The rodent responsible then takes another fruit instead of going to the ground for the fallen one. Later I or the dog picks it up from the ground. Such fruit are hardly damaged, usually with only a few teeth marks. Avocados have the ability to close around a damaged spot on the skin rather than rotting as most fruit do.

I keep many animals as pets at Hilltop also. One of the first were the koi who live in the concrete pond at the back of the house. Some of these multi-colored carp are over two feet long. The fish live from pond-fish food, cubed avocados and anything they can grub up in the pond. There is an aviary in the house, between the master bedroom and the back sun room. I had it put in when I had the master bedroom enlarged. It is peopled by two kinds of doves, midget quail and finches. The doves are constantly hatching young ones. In the meantime they give me pleasure with their soft low calls. There is also a small flock of chickens outside, penned because predators wiped out previous flocks that were given outside run. There is also a tank of salt water fish and there used to be one of turtles in the house, the fish for me, the turtles for Jessica.

Another form of life affirmation is growing plants. One of the most brilliant user discoveries of man was the domestication of plants and animals. Basically, he converted creatures that could be used for food and other purposes into productive slaves by learning to control their life and breeding cycles. Domesticated plants and animals became the basis of all civilizations and were incorporated ritually into many traditional religions, with the exception of the JCI Triad. The man-centeredness of those creeds kept practitioners from doing anything ritually except to sacrifice animals.

Anyway, although the Real Life Church emphasizes life

affirmation, it recognizes that life use must also take place. Indeed it accepts that during most of man's history, the hunting and gathering stage, wild plants and animals were the means of sustenance and thus survival. Thus it was natural that man, the domesticator, would use (eat) the creatures he had transformed. It couldn't be any other way. And so, though both flesh and plant foods are appropriate for man, it has happened that when life affirmation became established in some traditional religions, as well as in the Real Life Church, that a kind of repugnance set in for the continued consumption of flesh food.

It is obvious that man has to eat something which has been living, and this leaves only two choices. So if one type was to be discontinued to foster the life force, it would most logically be meat. Plants can do without animals, but animals cannot do without plants. It is a difficult distinction to make, and those of us that do not eat meats of many kinds for reasons other than health frequently have a problem explaining our choice. I used to tell anyone who asked that I chose not to eat those creatures which struggled not to be killed, which were of course the animals. And it must be remembered that even strict vegetarians have to eat something that must be killed. I suppose if there was some non-life food, some vegetarians might choose it. But to me plants were the least traumatic to kill and eat. I now no longer have to fend off aggressive questioners, since when asked, I simply say that I am a devotee of the Church. And generally Americans will accept odd behavior if they are told it comes from a religious creed. Anyway, I became a partial vegetarian and now know it was part of my initiation into the Church. But I must emphasize that one can be a devotee and still eat meat. A side effect on me of vegetarianism is that I became deeply involved in growing plants. Apart from the nutritive value of veggies, and their contribution to husbandry, growing them for consumption is life affirming. One of my greatest joys is cultivation, nurturing a seed until it matures into a fully productive plant, then harvesting and eating it. For fifteen years I have had a variety of vegetable gardens, plus 40 or more kinds of fruit trees. Thus the various acts of worship that take place throughout the year make me feel like a very holy man at Hilltop.

I indicated early on that the basic action of the life force is reproduction. And the basic act to make that happen is sexual intercourse. What makes the living world go around is not money, it is fucking. The simplest creatures get by through subdivision, splitting into two identical halves. But all complex creatures, the ones we are most familiar with, from Mendel's sweet peas to *Homo sapiens*, depend on members of the opposite sex to recombine genes, to make a different other. So it should come as no surprise that sexual intercourse is ritualized in the Real Life Church. Having sex is an act of worship.

Sometimes at Hilltop sexual intercourse is obvious, at other times it can only be deduced. The koi go crazy in late spring when the water temperature rises. They slosh around for 4-5 days in groups of one female and 3-4 males. The males punch the females to get them to release their eggs, after which they eject their sperm onto the precious globules. The male doves, both inside and out, go through a prance dance to get the females in the mood. I have seen hawks careening wildly around the sky, the males trying to titillate the females. Once I had an invasion of gopher snakes in the garage, twisting and turning madly in the radiator of my pick-up, coupling periodically. All creatures, great and small, do it, as they must. And though a damper has been put on this primary act of life by man, and especially in the JCI-Triad, it still goes on with him/her, even if in private. We celebrate the process at Hilltop, considering it the chief miracle.

Some might believe that most of what I've discussed can only take place in the country. That is, that a devotee has to live in a rural place; whereas most people nowadays live in the city. Few in the developed nations are country animals. Even I have been a part-time countryman for only a few years. But I wish to emphasize that no matter how different it seems, the city too is a part of nature. It is wrong to think that the city is separate from nature. The megalopolis, with all its complexities, is just as natural as the simple hunting and gathering band was. Evolution created both. And rituals and miracles take place at all levels of complexity. One can grow plants in ones yard. One of the best vegetable gardens I ever had was in my front yard in Westwood, a very toney section of Los Angeles. Creatures as pets are more available in the city than in a rural area. There are even wild creatures of considerable abundance in the city. I had wild birds in all the city places I ever lived in. And there is now a whole new type of wild creature in the city, the immigrant from the rural areas, the opossum, raccoon, skunk, coyote, crow, etc., all survivors who have learned that the city is a great place to raise a family. All are life forms which can he worshiped by any devotee. As for the ultimate rite of the Real Life Church, sexual intercourse, it can take place anywhere, and most often does take place in the city.

21

A Tall Tale From the West

When I was teaching about language in introductory anthropology classes I used to try to get participation from students by asking the rhetorical question, "What is the function of language?"

The first response was always, "Communication."

That was what the students had heard in pre-college classes and in its own way it was basically correct. Unfortunately it wasn't very helpful, and certainly did not point in the direction I wanted to go. So to steer them more precisely, I would say, "Well yes, if you mean transmitting messages from one person to another. But if you mean passing true information about actual occurrences, I'm afraid I'll have to differ. Of course it is true that we use language as the primary vehicle for passing on the information from one generation to the next of what we call our culture, our enculturation. And certainly most of that needs to be true information if the cultural system is to continue." Nods, note taking or dozing. I would go on. "But language also serves as a vehicle for getting people in a social system to cooperate. It is a binding force for society."

Pause, then hopefully the grabber. "But that doesn't mean what is said should be true. In fact, what is said might best be untrue if the purpose is to get people to work together. It is well known that political leaders frequently do not tell their constituents the truth, particularly if they need compliance in situations that are fraught with risk. Warfare is a good example. The citizens are called on to risk their lives. How else to get such sacrifice better than by withholding frightening information or telling untruths? Moreover, to get war fever going it is very useful to blame despicable things on the other side, true or untrue. The Vietnam war was built on an edifice of untruths, from the Bay of Tonkin incident to the body count of American soldiers and how many soldiers on the other side we were killing. Diplomacy has long been accepted as based on withheld or false information, from country to country, and from leaders to citizens. For other countries we call it "the big lie," but name me the country which doesn't practice it. New recruits learn how to tell untruths when they learn the ropes of diplomacy. Thus it is perfectly logical to get an Ollie North and an Iran-Contra scandal. And who can blame him as an individual? He learned the noble art of deception from the old hands.

Now if we accept that politicians and diplomats are deceitful, must we get rid of them to get back to a morally clean state of affairs? I doubt it, unless we are prepared to do without culture overall, though I do think there are more things that might be done to keep official lying in check.

Because let's face it, telling untruths occurs throughout the social body, not only in government circles. I am by no means the first person to observe this fact. I still remember the piece on lying by Mark Twain, who claimed that he started lying even before he had learned to talk. He claimed, tongue in cheek, that he was stuck by a diaper pin one day and cried. Whereupon someone came in and picked him up, fixed the pin, then rocked him to get him to stop crying. Whereupon the next time he got bored, one of his regular problems in the cradle, he claimed, he remembered how the person had come to hold him when he cried. And so he did it again, even though no pin was involved. But he claimed he continued lying the rest of his life. And he could do it easily, even without speaking.

When he grew up, he would tip his hat to all kinds of people he didn't like.

Children lie frequently. What else can they do when they are interrogated by powerful adults? I know I did, about the books I read, where I went, and the windows I peeked into. These were all sins and when I went to Catholic confession, the number of lies I told, usually to my mother, was a regular item. As I recall, the priest usually wasn't very interested and simply lumped my lies in with my other sins, like bad thoughts and skipping church services. He would just add in a few extra Our Fathers and Hail Marys for me to say as penance.

During my field work days, I early came to the conclusion that peasant villagers regularly lie or deliberately withhold information from powerful people, either those of their own system or powerful outsiders. A number of anthropologists have noticed this characteristic. An anthropologist by the name of Berreman described how some of the people he had interviewed had told him untruths when he had been doing his study in their village. They had fear of powerful people in their own society (Behind Many Masks) who might take reprisals. More recently the anthropologist, Chagnon, claimed that "his people," the Yanomamo, lied consistently about their genealogies, the best I remember as a joke.

Years ago an ethnographic mystery novel, "Call the Next Witness" impressed me very much. Despite its title, the book was about Indian villagers. A death had occurred and the chief constable, who was also an Indian, interviewed one villager after another who might have been involved. The investigator had one over-riding premise -- that all villagers would be lying. However the lies of each would be in the direction of his vested interest. Knowing village mores, the constable was able to guess the vested interests of each according to his social position. Thus, he narrowed down the suspects until the guilty person was exposed. If children lie because they are powerless and exposed, why wouldn't it be natural for villagers and tribesmen to lie when they are exposed to powerful outsiders, whether bureaucrats or inquisitive anthropologists? Which brings me to another social type which must lie for self-protection, criminals or others who are hassled by police. Lacking power, at least when they are hauled in to the police station, or confronted under duress, they can only lie or withhold information to try to save themselves.

That is why when there is an incident in an ethnic neighborhood and the police intervene no-one has seen anything.

I would have orated all of the above to my anthropology class and sum it up by "And though one of the functions of language is to pass on information, the primary social function is to influence others. If passing on information would serve the social function, so much to the good. But it may well be that in a given situation deception may be more useful than honesty."

I used to try to bring out points in my lectures by using incidents that were of interest to the students, rather than leaning on examples that were remote from their experience. And since college students, and certainly mine, were normally young adults, I used this example, "For instance, take the dating game. What is the primary need to decide before giving or asking for the other's phone number?" And without any preliminary back and forth, the students' guesses as to what I was looking for and me trying to steer them in the direction I wanted, I would say, "It is to try to sort out what is being told, to separate reality from fancy, truth from lies. For instance, if he tells you he is an electronics engineer, could he really be a TV repairman? And if she says she studies psychology all the time, does this mean she spends a large part of her time watching soaps and reading tabloids?

"One thing is for sure, and that is that he or she is not about to tell you anything that would damage their chances -- if interested. Later on, you will find out, but by that time you might be committed."

Generally by this time I would have got the message across, at least partially, which is all one ever hopes for in a class lecture. But sometimes it wasn't so easy. In one class I remember, I really got carried away trying to get this point across. It was a very responsive class which had really turned me on also. A totally new gambit occurred to me. I said, "And since sorting fact from fiction is a vital necessity for us humans throughout life, we will start here. I will give you some practice. Some of my statements will be true, others false; and I will not tell you which is which."

That grabbed their attention. There was a lot of shifting in the seats but quietness otherwise. Finally one spoke out, "But how will we take notes? How will we know what to study for the examination?"

A Tall Tale From the West

"You won't, unless you are good at separating truth from falsehood." I added, "And after all, you will be learning. And that is what you are in here for."

I was wondering by this time if this little game was a good idea but told myself that at least it had brought the group to life, no-one was dozing. So I went on to discuss other characteristics of language. The students started to jot down notes but then I noticed a rapid slow down. One would peer over at another and then whisper. One put up his hand. I stopped, "Yes?"

"Professor Niehoff, was that last statement true or false?"

I just couldn't help myself when I said, "I was sort of expecting that question but unfortunately according the rules of the game, it cannot be answered. Because if I said it was true, I might be lying, and if I said it was false, I might be telling the truth. You could never know from asking me."

"But how will we know what to study for the exam?" They were seriously perturbed. I figured I had gone too far. So I informed them that the game was over. I had simply used that small bit to get an idea across that I thought was valid. But I also knew note taking wasn't easy, they had to pass their exams, and I had no right to make it harder on them. Things settled down and we continued for the rest of the class quite satisfactorily.

So what is the social function of language? Generally it is to get the listener to do something we want. The State Department or military public relations man who passes out bullshit at news briefings day after day is trying to manipulate the public to accept his war as legitimate. This is to encourage their sons to go off to fight as well as to support the effort financially when more taxes are imposed. The child lies to his parents so they will not cut down his allowance, or "ground him." The inner city ethnic and villager clams up or tells falsehoods to the cop or ethnographer to get them off his back, or at least not give them ammunition to be used against him.

But there is another function of lying. The biggest lie I ever told, which I am leading up to, would make no sense if there was not this other function, which for right now I will label as entertainment, or giving pleasure. Many cultural systems have inculcated a guilt feeling for telling a lie so deeply in most people that they have a lot of bad feeling from telling someone else a deliberate falsehood about a serious matter. And before someone jumps on my back indignantly about all the con artists operating out there whose success depends on deliberately telling falsehoods for fame or profit, the wheeler-dealers in international diplomacy and military exploits, the junk bond kings and real estate developers, not to mention the types who develop systems to gather up funds from newly divorced, inexperienced females and retired people, I will insist that they are not typical. In fact, the only reason their scams work is because there is a large majority of people who will believe an official and who will put their money into investments promising pie in the sky. I believe that most people, at least in American society, find it hard to lie for fame or profit.

But what about fun? Ah, that's another story. Orson Welles did a famous radio broadcast when he was young, an account of a Martian invasion in the U.S. It was done well enough that it scared the bejeesus out of a lot of people. And it was all made up. Now why, we may ask, would a radio producer tell such a preposterous lie? We can deduce that he did it for personal aggrandizement, to make money, or to make himself more popular, which could add up to more money down the line. No argument. But why such a wild lie? I will argue that we play with language, and that one satisfactory way to play is to exaggerate or create. And a wild lie gives pleasure in itself. Much more recently a report came out on national TV that Lenin had gone down so much in the estimation of Russians that his remains were to be disposed of by auction. The next day it was admitted to be a hoax, that no matter how much the Communist leadership had gone down, that the idea of auctioning Lenin's remains was made up by the TV announcer. Again one can claim that TV news announcers will do all kinds of things to get an audience, to build up their ratings. But why such a preposterous lie about a former national hero? It so over-ran the bounds of media propriety that pressure was brought to bear, probably by advertisers, to have the commentator apologize. Why the guy did it in the first place is because telling tall tales is a normal human characteristic, if not by everybody, at least by quite a few. There is even folk wisdom that man is a teller of tales. I once told my former professor, Arensberg, the story which I am about to relate here, and his comment was, "Ah, the tall tale of the West still lives on."

He was referring to the belief that in the old West, before the era of mass, electronic entertainment, that men told wild tales, traditionally around the campfire. Insofar as this was true, and I suspect it was, the teller of tales around the campfire was continuing a long tradition, that of the bard. Among other things, the tall tale was prime entertainment. And lying, either by exaggeration or out-and-out creation, made it much better. Basically there was no other motivation for either the teller or listener than to have pleasure.

When I was a young father, I had an experience how such acts of creation come into being. My wife and I would take turns putting our son to bed. When she did it, she ordinarily read to him. I suppose I just talked, perhaps discussing what he and I had done through the day. Anyway, one night neither of us seemed to have much to report, so after a bit he asked me to tell him a story. I presume I had done such before, though that is another of the innumerable details that have faded in my memory.

I must defend myself at this point by denying that this is mental deterioration from aging. Rather it is a consequence of living a long full life with a large accumulation of experiences. The old have so much to forget, the young have so little. No wonder the young have less memory loss.

Anyway back to the account of Turkey-Lurkey. My son asked me to tell him a story about animals because I was so good at that. So evidently I had told such stories before. I was tired but he kept insisting, without nagging, and I finally agreed to tell him about a turkey which I would call Turkey-Lurkey. The only reason I can remember I had decided on this was that Thanksgiving had just passed, when we had had a traditional turkey dinner. But in the style of animal stories to children in America, my story had little to do with what we actually do to domestic animals, eat them. Instead I made my character into a turkey-person and sent him off on a series of adventures. So night after night I told my son of adventure after adventure, as Turkey-Lurkey traveled through the Great Valley. And insofar as this essay is concerned, it was pure fabrication, invented to entertain a little kid. I suspect that many of the great folktales of the world got started something like this -- though usually to include adults. My story was totally untrue, a figment for entertainment.

Another characteristic of verbal deceit is the distinction

between telling falsehoods and simply withholding information. Withholding generally is taken to be a lesser evil, though still a minor offense. The policeman cannot prosecute the inner city resident because he "didn't see nuttin," though it must be exasperating. Withholding information is considered more than an impropriety in some legal procedures. A person can be prosecuted for with-holding that which is classified as against the law. And though one can't be forced to volunteer information in a legal procedure, the individual can be forced to answer "yes" or "no."

Still, in non-legal events, we think of withholding information as less heinous than deliberately telling a falsehood. When I used to go to confession and report the number of lies, I'm sure I never including occurrences of withheld information. We did have a distinction in the Catholic Church then between little sins (venial) and big sins (mortal).

Presumably when I was brain-washed as a child, no sin was specified for withholding information. And so I grew up with a strong guilt feeling for telling someone a deliberate falsehood for personal benefit, but much less for withholding information. As described in another essay of this series (Dealing at the Convention), I withheld the fact that I had failed my Ph.D. exam from my first employer and felt a little guilty about it, but much less than if I had told an out-and-out lie about it. I consider myself to be relatively well brain-washed (indoctrinated, enculturated, socialized) for American society. Presumably there are individuals who have no such bad feeling, but even of the con artists and government officials who lie a lot, many do not feel quite right about telling real lies. However, they must accept it as a necessity, and usually they have a justification ready. Presumably official procedures to force people to tell the truth are a consequence of the need to control the populace. How can you run an income tax system if people are permitted to lie? So you force them to sign their name, with the understanding that the powers of the state will come down if they have lied. But of course many do lie anyway, and a few do suffer from the powers of the state, though most probably do not get caught. Anyway a modern state is run by means of people being forced to do all kinds of undesirable things, like paying taxes, going to war, and supporting all kinds of programs they as individuals do not favor. It would be easy for the state if most individuals were not racists, did not mind risking their lives in wars about which they know little, and thought paying taxes were good things to do. But since they do not, the state forces compliance and sets up penalties for the non-compliers, based on forced truthfulness. One is not permitted to lie on a draft registration or income tax form.

Presumably religious systems also condemn lying because they need to manipulate the populace. Thus it is worthwhile to be able to force truthfulness or at least control lying. How could the priest deal with his parishioners if they lied at will. So lieing is a sin and the kiddo is made to feel guilty when very young, through his parents, who were indoctrinated when they were children. Then the priest tries to reinforce the truthfulness in his sermons, and when the parishioner goes to confession. It really works. Now 50 years after I dropped out of the church, I still feel guilty if I tell a deliberate falsehood for personal benefit. This is despite the fact that I can write dispassionately about the topic, and have told some whoppers for entertainment. And so on to the entertainment part of this essay.

One afternoon I passed through the audiovisual department of my university, accompanied by a colleague. I had been friendly for years with the fellow behind the counter. And as I handed him the film I had just finished using in class, he said, "This fellow, Niehaus, he any relation to you?"

My name is spelled Niehoff, but John (pseudonym) was not a bosom buddy, and he may not have known the exact spelling of my name. He was a fairly straight guy, and like a lot of others, seemed to think of anthropologists as pretty bizarre (probably true). Anyway I said, "Why, what about him?

"Oh he was on the news the other night. Got kidnapped by the guerillas in Venezuela."

Then I remembered having heard of this American oil executive who had been carried off by the dissidents of that country. He was no relation to me, probably not even of the same national origin. Even so, I started a lie. I said, "Oh yes, he's a second cousin, but he's alright." I seem to have the tendency that once I tell an initial falsehood, I make up other "facts" to support it. And so when he said, "Well, aren't you worried?" I replied, "Oh no, he can take care of himself. Why he used to be a secret agent, and he would disappear for months at a time. But he could take care of himself. If he didn't surface for a couple of years, we would get concerned."

John looked at me as if impressed. He said, "Oh, glad to hear that. You will let me know when he surfaces, won't you?"

"Surely, but don't worry."

I could see my fellow professor eyeing me sideways, and when we had got outside, he said, "You know, we know you and it's okay to tell us wild stories. We understand. But you can't go around telling them to others. Like John, he believed that cock-and-bull tale."

My response, "Well so what, what was the harm? He can go home tonight and tell his buddies or family what cool guys they have on the faculty. Their cousins get captured and do they panic? Oh no, they have full confidence in their relatives, this one being a former secret agent anyway. Now he's got a good story to take home. What's wrong with that?"

"But he'll ask you about the guy. Then what?"

"Oh, he probably won't. He'll forget about it. But even if he does remember before two years are up, I can tell him no news yet, but there's still plenty of time. And I doubt seriously that he will still remember it two years from now."

John never brought up the subject again.

But the grandest tall tale, and the one I have been leading up to, also took place in California. It is the following. In California there are some earthquakes. Just a few years after I settled in the state, there was a good-sized one, the Sylmar Earthquake. It occurred early in the morning and my wife, son and I ran out of the house. No damage occurred and we went back in shortly. There was a fair amount of damage elsewhere, and the quake got national attention, as events in California tend to. Like most other Californians, we quickly went back to our normal routine.

One evening, about 2-3 days later, I was home alone when the telephone rang. It was a former high school friend of my son, who was a telephone operator in Boston. She said she was permitted to use the phone when they weren't busy. To her request for Justin, I said, "He's not here right now and he probably won't be back until late." I said.

"How is he?"

"Oh, he's alright."

"I mean there was no problem from the earthquake, was

A Tall Tale From the West

there?"

I knew that many people from other parts of the country had called after the quake to find out about their California relative or friend. We who lived in California joked about it, the damage had been so scattered. So to pull her leg, I started on a story which got more and more complicated and harder to extricate myself from. It also kept getting more and more preposterous, so much so that I kept thinking she would stop me at any moment. I first said, "Well you know that when we get earthquakes in California, crevasses open up in the ground. And in this instance, one opened up right out in the front yard. And when we ran out of the house, Justin was first and he fell in."

Now this beginning was ridiculous. And though I can't say that no-one ever fell into an earthquake crevasse, I never heard of such, certainly not in California. The earthquake damage in California was almost always from buildings collapsing, and fires. Anyway I thought she would recognize the fancifulness of my story's beginning and call me to a halt. But evidently she was from that great mass of people, probably in any population, who tend to be believers, no matter how incredulous a story. Also I have the ability to tell fanciful falsehoods in a very straightforward way, and many people are taken in more by the manner of telling a story than its contents. Anyway there was quite a pause then, presumably her considering the story. Then she said, "So what happened? Did he get hurt?"

So I went on, "No, the fall didn't hurt him?" I added, "At least that's what he told us."

"Where is he now?"

"Oh, he's still there, at the bottom of the crevasse." Surely she would stop me. This was beyond the bounds of belief. But she didn't and so we went on to the very end.

She, her tone raised, "But what are you doing to get him out."

"Well, there's this overhang which he rolled under when he fell, and we can't see him. But his voice is okay and we talk and he's alright." I added, "We have lowered blankets and food and water on a rope and he says he's okay."

This was a lot for her to swallow and there was a particularly long pause, but she evidently did swallow it. The pause was so long though that I added to reassure her, "We're doing everything we can to get him out, but it might take a little more time."

Finally, "Is there anything I can do?"

It certainly must have been the devil in me when I said, "You can say some prayers if you want."

Another pause and then, "Is it alright if I call back, you know, to find out if anything happened?"

"Oh sure, call anytime." She hung up.

I went out that night on my own. My wife and I were mostly separated and she also had gone out on her own. I forgot about the call. But it wasn't over. The next morning when I was shaving to go off to school, my son came into the bathroom. He was very serious when he said, "Art, you can tell wild stories like you did last night to people who know you. But you can't tell them to strangers. They can't tell when you're lying."

"How did you know about it?"

"The police were here. They came to the front door, checking on a kid in a ditch. I presume they thought it was an accident or homicide. She called them after she hung up on you."

"So you answered the door and told them you were the kid, eh?"

"Yep." He stood there shaking his head.

His old school buddy had obviously not accepted the story totally. Otherwise she would never have called the police. Nothing further happened. But now when I think of language, I accept it as a means of transmitting information but also a medium for deceit, sometimes with serious intent and sometimes for the sheer joy of playing with words -- and people. And to put it into a folklorish mode, that is how deceivers communicate with believers and how tall tales are told in the West.

22

The Stroke

Another older man dressed in denims came out of the inner sanctum and went to the front of the secretary's desk, probably to settle his account. He was also probably a retired person, I thought. After all, this was a medical center in Fallbrook, the self-proclaimed avocado capital of the world. I knew that there were a considerable number who had come here from Los Angeles when they had retired from stock brokerage or medicine. I was not retired yet though I shared many of the same characteristics with these well-to-do people who lived in ranch style houses on the tops of hills, often surrounded by avocado groves.

I took the medical history form from the secretary and sat down. But before I began checking the boxes of the list of illnesses, I paused to consider where I was in life, and what else might be in store for me. Life had been good, though except for the stroke and divorce, the most recent years had been the best. Despite many problems, I had gained a great appreciation for life. I had learned to appreciate all life forms. It pleased me to think that I had long given up hunting and fishing and had become a vegetarian. My only regret was for the animals I had killed in my wanton youth. It only helped a little to remember that most of the other boys I grew up with had done the same. I thought then that if I were ever to go back to practicing a traditional religion, it would have to be Therevada Buddhism or Jainism, both of which I had learned about through anthropological studies.

In my youth also I had acquired a love of reading and writing which had served me well, both for maintaining my health and for professional advancement. I should do another essay, I thought, on "Writing as Therapy." And then came the great love of my cognitive life, the field of study of humankind, which had given me most ideas that I had. From my earliest college days, it had become my lens of perception. For many years I had been unable to consider any of the whirling occurrences of existence without applying some anthropological insight. Without those, the world would probably still be as incomprehensible as it had been in my mostly wasted teens. For decades it had been impossible for me not to take my work home. My work, anthropology, was my life.

I stirred myself to shake off the memory images, at least enough to answer the medical questions. "It will be just a few minutes," the nurse/receptionist said as she took the filled out form.

I sat down again and thumbed briefly through a <u>Sunset</u> magazine. But my mind would still not settle down. Almost without awareness, I ceased turning the pages or looking at the colored desert scene where I had stopped. Instead, I turned back in my mind's eye to Hilltop, my place in the country. That had been the most recent addition to my life, only two years before, following the two great traumas of my later years, the stroke and divorce. The purchase of the small grove, mostly planted in avocado trees, had come from my need to foster life, plant and animal, domesticated and wild, as well as providing me with an unlimited supply of my favorite fruit and a place to retire when my teaching days would be over. So far, the farm had been a total pleasure, even with the weekly commute of 100 miles each way to Los Angeles.

"Dr. Niehoff, you can come in now," the lady in the uniform said, opening the door to the inner sanctum.

I followed her to my appointed cubicle, where she took my temperature, pulse and blood pressure, marking each on the chart. "'The doctor will be here in just a few minutes."

She closed the door and I was left alone, sitting on the edge of

The Stroke

the examining table, surrounded by the antiseptic accouterments of 20th century medicine. Again my mind wandered, though this time following another bent. I began to project my thoughts into the future. I had learned in anthropology of this human proclivity. They called it "time binding," the ability to look backward and forward, to survey one's history and imagine what might yet be. Anyway, there I was, trying to think forward, trying to peer through the mists of the future, trying to see what might yet happen. I knew that though I had been far more fortunate than most in so many ways, that there was still one large part missing in the web of life. My divorce of three years before had taken away one element of living that I had always seemed to need, a feminine companion. Some men may be able to do without a woman, and vice versa, counterbalancing the loss of the pleasures with the negative of some lost independence. I had never been such. Although a love of life, reading and writing, anthropology, and Hilltop had all brought much fulfillment, I still could not be completely at ease without a feminine counterpart.

The mists swirled, thinning, then thickening, lightening, then darkening, and I sat mesmerized, watching the shadows and forms as if something of moment were to occur. And a face began to take shape, round with high forehead, full cheeks, gentle brown eyes, with a small nose and lips. Auburn wavy hair, almost shoulder length, surrounded the face. It had to be an attractive, understanding woman. How could I have known that the lovely phantom was a foreview of Jessica, the love yet to come in my 60's.

The door opened and a tall man in his 50's, bespectacled and wearing the traditional white smock, came into the room. It was Dr. Seebring, my new physician. We shook hands and he studied the form I had just filled out. Then he did a short physical exam. "Your condition seems quite good for your age." I was sixty-one.

"You know you made a remarkable recovery."

"I have been told that several times. I guess I'm lucky."

"You seem to keep yourself in good condition."

"I try." I did eat moderately, vegetarian, I did not smoke, and drank moderately. Also I kept up my exercises, riding my ten-speed, jogging and playing tennis. And since I had bought the avocado farm, I spent a considerable amount of time going up and down hill, tending the grove."What should I do for maintaining my health?"

"I think continuing as you have will be fine."

I was on light medication for my blood pressure, which had been normal for three years. Otherwise I kept busy as a farmer and college teacher. My life had stabilized since that morning three years before when I had awakened with a violent headache, which had rapidly escalated into a stroke with paralysis of the right side. The recovery had not been easy, and as a consequence of that illness, my life had changed considerably. But I had recovered from one of the main killers or debilitators of middle and old age, to mostly resume the kind of life I had before.

How had it happened? In looking back, I see a three year period in which first came the stroke, seemingly from nowhere, then a tortuous five months in which I often felt that recovery would never come, and finally the two and a half year's climb back to normalcy. There were climax points that I will not forget.

It happened early one morning in February after a good night with a group of anthropologists. I had been the main speaker at an anthropology club meeting, discussing the problems of fieldwork. I had done more such than anyone else in the department and felt that those experiences had been some of the most important highs of my life. After the meeting, Carla, my wife, and I had returned home where we visited with one of my long- time colleagues who had come to Los Angeles that night. When we went to bed at one o'clock I felt tired but good. The stroke came at five. After sleeping peacefully for several hours, I woke up with a headache. At first I thought it had been caused by sleeping with my neck crooked. But no matter what I did, it got worse, and within an hour I was vomiting, the room was spinning, my right side was paralyzed, and a hurried call by my wife had produced a diagnosis of possible stroke. Some part of my brain was blocked or burst open. My wife and the friend who had stayed at the house that night got me to the nearest hospital where the medicos in emergency worked on me for eight hours.

My recollection of that night and morning was of three or four people busily trying to do things to me. Their faces were blurs, and though I tried to do whatever they asked, my thoughts were elsewhere. Despite my shock and continued disbelief, I was turning inward, trying to marshal whatever energy I had left. I would force the paralysis out of my limbs and face. Nothing complex, simply

The Stroke

focusing my remaining energies.

The one person I remember most vividly was a very concerned middle-aged nurse. She inspired my confidence and I especially tried to do what she asked. I had gagged each time the intern had tried to get a tube down my nose and into my stomach. The nurse then entered my focus, saying, "I know it's hard but you must try not to gag. It's for your own good."

I tried especially hard for her and the tube went into place.

Many months later I returned to the hospital, walking upright as a man should. I tried to find my way to the emergency care unit, wanting to thank my saviors. No one was there that I recognized. I walked away a little sadly, thinking that those people saved lives night after night as part of their job and probably only rarely found out what happened to the wracked bodies that had been delivered to them.

After a couple of days in intensive care, I was put into a regular hospital bed. My new doctor informed me that I was lucky that the stroke had taken place in the low brain and was of a kind from which good recovery could be expected. The important need then was to lower my blood pressure, something I had known nothing of before the stroke. I did remember though that high blood pressure had been common among the men of my father's family and that both my father and father's father had died from it.

My only memory of my grandfather was of him being put outside on warm days, my grandmother wiping the drool from his lips. She had a towel for this purpose. I unconsciously put my left hand to my still partially numb lips and was pleased to note that the fingers came back dry. I thought then that I was lucky indeed, much more fortunate than my forebears. I might well return to normalcy. I thought then of getting back again to my classes in anthropology, mounting my ten-speed, and enjoying the simpler pleasures such as strolling in the bright sunlight of a California day.

As soon as I was released from the hospital, I began walking, walking, walking. To keep in a straight line was not easy, because the stroke had affected my sense of balance. I was especially cautious not to fall down. It embarrassed me to think that others might see me as drunk. But I kept increasing my distance each day.

Some three weeks after the stroke I got out my ten-speed. I had not been sure about going on it because of my impaired balance, but Dr. Norman had given me the go-ahead. He said, "Since you are an experienced bicyclist, you ought to be able to handle it."

We were wrong. On the first ride, even though I was super careful, I quickly came to the opinion that I should not yet be out on the bike. I wobbled too much. Moreover, I could not turn my head backward without getting dizzy. I decided to return. Unfortunately, I decided to ride back.

The next thing I knew I was lying on the street at Wilshire and Sepulveda, writhing like a snake with a broken back, expecting at any moment for a car wheel to pass over my head. The line of traffic stopped and a circle of concerned faces materialized. "Are you alright?"

The good news was that it was only a broken thigh bone. I got out of the hospital within twenty-four hours, rods having been inserted down the middle of the bone, I supporting myself on a three-sided walker. I quickly became dejected.

My wife became dejected too. She started to have difficulties on her new job and failed to pass probation. She said her supervisor felt it was because of the stress from my illness. I did not take that seriously enough, being too concerned with my own recovery. We began to argue about all kinds of things. A few days later when she visited me, she announced that she had started psychotherapy. In the meantime I started to work on learning to walk with the stroller. First back and forth across the house, then carefully down the back stairs and return. I got better at this new form of locomotion, though I tired easily. In between stroller practice I would lie on the couch. I had no appetite. Getting food was too much trouble, so I spent most of my time sleeping. When the phone would ring, I would delay as long as possible, hoping the caller would hang up.

I soon noticed that my pants were getting looser. But that wasn't enough to shake me from my lethargy. When my wife and colleague took me on a regular visit to Dr. Norman, he was quite concerned. I had lost almost forty pounds, which was a lot because I had not been heavy. Dr. Norman felt I should go back into the hospital because I was suffering from depression. I couldn't believe it. It seemed impossible that I could survive another stay. They would put me back on intra-venous feeding of sugar water and I would lie there lumpishly, to come out no better than before. I turned to my wife and friend. They both agreed that I looked bad. As we drove to the hospital the third time, my wife and I broke into tears.

The worst I remember of that stay was how terribly constipated I became. I had practically no appetite to begin with and found the hospital food very unappetizing. Practically everything was overcooked, there was hardly any roughage, and salt had been eliminated from my diet. I have been a vegetarian for many years so did not take meat dishes. The vegetables they served were very insipid. I hardly touched them. And though I have had very regular bowel movements most of my life, I stopped going in the hospital. After more than a week, they gave me enemas, which were no help at all. I remember being awake during most of the nights, embarrassed because I couldn't control myself any longer. Although I defecated very little, it was enough to stain the sheets. I kept a roll of toilet paper and spent several nights mostly wiping myself .

I did not really get much better, though I forced myself to eat a little more, and after a couple of weeks, convinced the doctor to release me. His one special condition was that I go to a therapist when l got out. My wife had gone to a religious retreat for several days so I got a friend to take me home.

I remember how joyous I was going away from the hospital. I was determined that this time I would make it, no matter that I looked like I had just been released from Auschwitz, I had not gone to the toilet for ten days, I had no energy, and my relations with my wife had continued to deteriorate. But still I was determined.

Dr. Seebring sat at a typewriter, composing an account of my condition. When finished, he gave me a copy and said, "You seem to be doing quite well as you are. So, simply continue as you have been." I headed toward the receptionist's counter to sign the necessary papers.

There are several occurrences which stand out as high points in my recovery. First, I learned to eat again. My wife had ceased to prepare anything due to my lack of response. Fortunately, I had learned as a boy to prepare food in my mother's kitchen. And so I did again in my house of illness, though it wasn't easy. I had to ease myself into a chair in the middle of the kitchen, then to prop open the refrigerator door so I could find what might be nutritious and palatable. Eating became a ritual in which I almost felt the spirit of my long gone mother was participating. It seemed she would say, "Now son, go to the refrigerator, get out some cheese and bread, or something else good for you, put it in a plate, pour yourself a glass of milk, and eat. Thus, you will get well, my son." Also, a fellow anthropology professor used to call me and refuse to talk until I went to the refrigerator for a bowl of ice cream or some other food. He would then tell me news of the department while I swallowed spoonsful of ice cream. I was exultant when the weight began to come back, two to three pounds per week. In between, I spent hours on the toilet, trying to get regular again. I was worried at first that I would get hemmorrhoids, that my anal sphincter would lose control, but the diet with roughage soon paid off and I got back to my regular schedule of once per day.

Another event of significance was the revival of the tomatoes. I have been an avid gardener for years, one of my basic life rituals. I had bought some patio tomatoes to plant in February when I was on the upswing. The broken leg had stopped that. The tomatoes had remained on the picnic table in the patio, being watered sporadically by my wife. They had grown in spurts, drying out in between. Not surprisingly, they looked miserable, straggly with dried leaves. But each had a tip of green. They had gone through a hard time, as I had, but they too were hanging on. Life was not to be abandoned easily. As my energy began to return, I decided to help them. I would bring them back to health and growth, as I was bringing myself back. I began to water them regularly and was overjoyed when the green tips began to lengthen. As soon as I could, I planted them in the ground, bringing out a box to sit on because I still could not lower myself to the ground and get back up. They came into production six to eight weeks later. Then I got back on the ten-speed. There was an added difficulty because my leg had become twisted by the break and set. But I persevered, though far more cautiously than the first time. The milestones were first to pass the corner where I had fallen, then a ten-mile climb up to the top of Sepulveda Pass and finally a thirty-five mile climb over the top of Topanga.

About this time also I started back teaching anthropology again, beginning with one class. I was still not steady on my feet, and during the first lecture I learned to be cautious. I was going on about enculturation, proclaiming in my deliberately loud classroom voice,

The Stroke

"I think of it as primary brainwashing ."

And when I turned quickly to write the word, enculturation, on the blackboard, the ceiling began to turn. I got dizzy. My momentary fright was that it was the beginning of another stroke. I quickly put my hand on the narrow chalk trough below the board and steadied myself. I felt foolish as I stood there, attempting to recover my equilibrium. But when I finally turned back, slowly, the students were showing no reaction. It must have been briefer and lighter than it had seemed. I carefully walked to the podium and put my hand there for support, acting as if I was studying my notes. Slowly the world became steady again.

The dizziness, I soon realized, was a consequence of brain damage. Dr. Norman told me that I would probably have that weakness the rest of my life. Fortunately, I did not. It gradually went away as new synapses took over. Within six months I could again teach full time.

The final positive occurrence was the acquisition of Hilltop, my avocado farm. After two years of efforts to readjust to my newly divorced status, I bought five acres of country in San Diego's North County. This turned out to be the most delightful place I have lived in throughout my life. Here I have been in the midst of life, of animals both wild and domestic, and a variety of plants that I could not have imagined in my wildest dreams. Moreover, Hilltop has a spectacular view of the hill country of this semitropical corner of California.

As I walked out the door and across the lot to my pick-up, soaking in the heat of the midday sun, I was pleased that the door to that room of my life was closing rapidly. Of course I would remember, but most of all in my thoughts was that I had beat the odds, that there was yet a distance to go. My steps were light as I headed toward the little truck. How could I complain, I thought. My health came back totally. I am older now, but that is a condition of existence I have learned to live with. But when I think of what might have been, the spirit world intervenes. The ghost of my father looks at me and says, "Look at Arthur. He became what he always said he would when he was a boy, a farmer."

And the spirit of my mother says, "Look, there he is, my boy who learned how to cook and nursed himself back to health."

And if there were a godling of anthropology, he might say,

"Look at Arthur, he needs to stay longer. He is not yet finished passing on the message."

23

On Being a Latter Day Veteran

My 51st Memorial Day since military service has just passed. That is, I was discharged from the U.S. Army in 1945. And after reading the customary eulogies for veterans in the local newspapers once again, I decided I had better get on the stick. As the saying goes, "No one is getting any younger." Moreover, during the last ten years or so I have been trying to put down my version of this chaotic universe, for whatever it is worth. And about veteran-hood, what I remember is a far cry from what I see in the newspapers. My perception of war and military service is hardly typical, like my perception of many occurrences.

I believe my father's attitude toward war and military service was more normal. He was a veteran of World War I, wounded slightly by a piece of shrapnel, receiving a small pension thereof, and as best I remember, his military service was one of the most exciting events in his life. I remember that he would come home highly elated when he had met "an old war buddy." I was quite interested in learning what had actually happened when he was at the front. But he wasn't very talkative about it. The things I remember that he did tell me were that he was a member of a machine gun crew and that his job was to carry the barrel, and perhaps some ammunition. I feel sure he told me he never had to fire the gun. Another thing I do remember was that he had refused to take a job as a listener on the German telephone line. You see, he understood German well, having been raised in a German-American community in Indiana. Also I remember him talking about poison gas a few times, I believe in the context that this or that war buddy had got some. And that is all I remember that he told me. I remember nothing about the mud or rats of trench warfare or any other of the horrors of that conflict I learned about later. Oh yes, he did mention one of the artillery shells, that he called something like a whiz-bang, and that when the whiz stopped and until you heard the bang far off, that you really had to watch out.

My father lived a quiet life as a working man in Indianapolis and died in his early sixties of high blood pressure and strokes. His wartime experience had been positive enough that he had arranged to have a military funeral. When he was buried, a local military unit, which had been assigned for this purpose, fired a salvo in his honor and a bugler played taps, to be answered by another in the distance.

It was moving, even to me, though I had then not got much into reading real history and learning what a dreadful butchery World War I was. And with all the horror, it was one of the most useless of wars, perhaps as senseless as the Vietnamese War. Millions were killed, not to mention the cost, and in the long run no-one benefited. I find it ironic that the one "great" who thought his service in that war was a high point was Hitler. That veterans could be proud for having taken part in such an event saddens me. It also saddens me that if my father had lived longer, I would still never have been able to discuss the war with him because he, like most veterans, hardly studied the causes and consequences of war, but generally just accepted the national version. So in a sense I am writing this little piece because I want his story, which is probably standard, to be told.

And though my own story is different to the extent that a military funeral would make no sense, before I recount it, I want to mention another soldier's unfinished story. And in that case it is totally unfinished because no one, now or ever, will know how it turned out. That is the story of Hans' father, in the monstrous event spawned by my father's war, World War II. That was my war also, of which I will shortly give an account. But first let me give what I can of Hans' father's war to indicate that from the soldier's perspective, it hardly matters what side one is on. Because Hans' father was a German national. I knew Hans since his father was long gone. He disappeared at the battle of Stalingrad, along with hundreds of thousands of others, on both sides, not to mention civilians.

I met Hans a few years ago in Los Angeles. He is a chemist/administrator in his early 50's, of German birth but with American citizenship. He grew up in Stuttgart and still remembers the bombing raids. I early told him that though I had flown over the devastated cities of Germany, that I had been in the Troop Carrier Command and our cargo was mainly of wounded soldiers. I did not volunteer that we also carried paratroopers and pulled gliders full of armed troops, which would fight it out with German troops on the ground. Or that we were carrying gasoline to provision the tank units at the front, when they were racing across Germany toward the end of the war. I do not know that it would have made any difference to him, since he had long adjusted to living among the winners (us) who have of course proclaimed our glory in the media and elsewhere at great length. Anyway in one of our talks, since I had noticed that he often spoke of his mother, but never of his father, I asked about him. "Oh he never came back from Stalingrad," Hans said simply, as to indicate that there was nothing more to say.

"You mean you never heard anything?"

"No, nothing."

There were hundreds of thousands who never came back, who were never heard of again. It was as dreadful a battle as some the French and English had been in during WWI before the Americans joined the conflagration. I did not know quite what to say but couldn't quit yet. And since I knew Hans as a very open and honest person, I asked, "Was he a Nazi?"

Again as a matter-of-fact, "No, but all men his age had to go into the army."

"What kind of work did he do?"

"He was a mail carrier." He added, "He wasn't interested in politics."

I thought of him as having a life like my father. Both had comfortable lives in their respective countries, not concerned with international events. They were just ordinary persons who had been sucked into the two worst international conflagrations the world has known. They were innocents who had been told what to do and who to kill. Probably neither questioned the rightness or wrongness of their society. And either could have been on the other side except for a decision some ancestor had made generations past.

And what about me? Was I that different? Not really, except that I have lived a long life and spent much of it studying and thinking about social events, including war. It was perhaps inevitable that I would someday analyze my own past, as well as that of the ordinary soldier generally.

I did not question the right of the nation to demand that I serve as a combatant when my time came in 1940. I hardly remember that anyone did in the working class from which I came. And despite the propaganda stories and films they began to bombard us with, I recall no hatred of the enemy. Though the media has endlessly proclaimed the "day of infamy" speech of Franklin Roosevelt, and how this galvanized the nation, I remember nothing about the news and reaction to the bombing of Pearl Harbor. I am of course using my power of recollection as an index of the significance of events to me at the time. For instance I can remember certain things very clearly, others hardly at all. I even remember quite a number of images, such as those of nubile young ladies, but nothing about the upcoming war. In other words the primary significance to me was that my life would be disrupted.

It is quite common nowadays to emphasize how much effort, or even their lives, that veterans were prepared to give up. I must confess that I had no feeling that we as a nation were going to lose anything, or that I was about to go forth to defend family, nation or democracy. And I was certainly not prepared to give anything up, though it seemed inevitable that I would lose some time from my life.

Neither I nor practically any other ordinary folk had any idea of the Nazi concentration camps. And even so, I accepted as given the fact that I would go into military service. When I think back on it now, after 45 years studying the brainwashing we anthropologists call enculturation, I see my acceptance, and that of almost all other young men, as no more than a natural consequence of what we were told all our lives. We did what we were expected to do. We were just like an Aztec young man who went off willingly to a "flower war." He was expected to capture victims for sacrifice and he did so. Or we were just like an ancient Greek "hoplite" warrior who submitted himself to 30-60 minutes of sheer terror in a phalanx battle. A well functioning society instills into its people a willingness to participate in difficult, or even life-threatening events. Once a large proportion refuse to accept vital roles, that society, or at least that particular social pattern, is in jeopardy. War always implies risk for some and boredom for many, so when individuals decide the negatives outweigh the positives, the social pattern may change or the society may even disintegrate.

In the three instances described here, the young men all accepted their lot. And the armies of which they were a part were truly effective. I am not here discussing the goals of the respective societies but whether the soldiers were effective. I believe they were. And in this respect, it doesn't matter which side had "god on their side" or that one side lost and the other won. The German and Japanese soldiers were probably the most effective since the Napoleonic or Mongol Wars.

The enculturation (brain-washing) of young American men broke down during the Vietnamese War. Then very many (the majority?) decided independently whether they would fight or not. Many decided not to do it at all. They evaded the draft. And many who went did not fight with all their capability. They smoked pot or took other drugs to achieve a different mental reality. That the war itself made little sense is not the point. Have other wars in which the soldiers were very effective made more sense except that they were taking over somebody else's territory? The conquests of Alexander the Great, the battles of the conquistadors, the British taking the Indian empire, the Napoleonic Wars, and so forth. The effectiveness of soldiery seldom has anything to do with the rightness or wrongness of the war. Effectiveness of soldiers is of individuals who fight no matter the logic or justification of the war. I suspect my father never questioned America's participation in the First World War, and I have no reason to believe that Hans' father fought less because while not himself being a Nazi, he was fighting in a Nazi promoted war. And as I have reiterated, neither I nor the other young men I remembered were concerned with the causes of World War II. We simply did what we were told to do. And because of that it was an effective army.

But to get back to my account, which I am giving as being more typical than the accounts of heroic deeds that are given to us on Memorial Day. Because that is simply a later form of brain washing, of the kind that made us accept our participation as soldiers in the first place. When the war materialized, we young fellows adjusted to the fact that we would probably be going. There were signs everywhere, not the least of which we were all registered for the draft. Then it became a numbers game as to which branch of service we would enter. If we had any college education, we could try for officer's candidate school. That was more prestigious and one could have access to a higher class of girls. I had only one semester and was not doing well in my second, so that didn't seem likely for me. I was not enthralled at the idea of being in the ground forces or the killer elites. I did have two brothers-in-law who did enlist in the marines and got into some real bloody conflagrations in the Pacific. I didn't know them then so have no idea what prevailed upon them to go that route. Only one seemed to have been a real gung ho type, though from his middle age on he was so obese his heroic role then was hard to imagine.

Anyway, many of the rest of us made choices primarily to avoid ending up in the ground forces. My best friend at that time volunteered for the navy which was supposed to have had a better lifestyle (living on board a clean ship, eating in a mess and sleeping warm and dry), as well as traveling a lot.

At that time I did not think of the fact that the United States was usually raising hell on somebody else's territory or waters. I don't remember when they got around to calling our military the "defense department" but in truth it should be called the "offense department" because we almost never used it for defending ourselves. Instead, from the time we started displacing Indians until the Gulf War we have always invaded other peoples' territory. Oh, perhaps the War of 1812 could be considered a defense action, but it was a piddling affair compared to our serious wars.

Anyway back to how we were volunteering. One friend got into the officers' corps because he was in good graces in college. Of course there were more in the upper classes where the sons had easier access to college. I was jealous. So I decided to head for the "wild blue yonder," the Army Air Corps. It too gave one officer status, and besides one got to fly in those glorious flying machines. Another thing I never considered in those innocent days was what an indiscriminately destructive machine for war the airplane was. Probably no device for bringing an opponent misery since the burn and destroy methods of the Indian wars was the bomber.

Like many others, I was a sucker for the recruiting films of Jimmy Stewart. That too is a set of images I still remember clearly, the tall, lanky, low-keyed Stewart, clad in his officer's uniform, with his big, shiny officer's insignia on his crunched, billed cap, and his rank clearly indicated by the epaulet insignias. Anyway, he would come up to the camera in his laconic, easy manner and speak to us with the sincerity that made him a natural as "the all-American boy." I cannot remember what he said but I am sure it was some version of how bad the Nazis were and how we were going to settle it for good and all with our glorious flying machines. Little did he or I know then of the coming firestorms of Dresden and Hamburg, not to mention Hiroshima and Nagasaki.

In any event, young and inexperienced as I was then (the ideal state for a soldier in a modern mega-war), I came out of the movie house all jazzed up by the glories suggested. Also it was made clear in those days that one did very well with girls if one was a hero in the Army Air Corps, much better than "grunts" (ground troops). Also at that time I had had a couple of hours instruction in a Piper Cub, and flying was indeed exciting.

Motivations are usually complex and frequently multiple, and my decision to become a "flyboy" was no exception. Another motivation I had then stemmed from the fact that I was having a lot of trouble getting along with my parents. Also I wasn't doing well in college, which I had started part time after I got my first job. And since it had been my decision, I had to pay for it, which was not unusual in the working class. Anyway, I was looking for a way out of my dilemma. What better way than to enlist in the Army Air Corps, which I did forthwith.

Suffice it to say that I did not do so hot. I did get some pilot training in a checkered career in and out of pilot and radio operator's school, ultimately becoming a radio man. None of the glories I had imagined came to me from being in the Air Corps. I did finally get a couple of girlfriends, sequentially, and had the wildest sex experience of my life while in uniform but as an enlisted man.

It took me a long time to adjust to my failure as a pilot, though eventually I came to the point when I was truly happy that I had never dropped bombs on civilians or other enemies. And perhaps because of my failure then (I thought), I was bound and determined to go through college afterwards, to learn about this blooming, confusing life. Thus, I became an anthropologist, college professor and writer, more or less in that order. But most important, I became a thinker, which above all else has supported me in my old age.

When I was discharged, the option was to remain in the reserves. There was a slight remuneration for this and a requirement of a couple of weeks training a year. I remember that I decided very quickly that I wanted completely out. I think the primary reason was that army life had been very boring and completely controlled by an inexplicable bureaucracy. I wasn't then anti-military, merely individualistic. Thoreau would have understood my motivation well. The choice was very fortunate for me because only five years after the end of World War II, the Korean War broke out. Men who had signed up for the reserves were called back. And even though I was not in the reserves, I spent many a sleepless night imagining that they would institute a new draft; and I would be in uniform again.

Which brings me to another point, uniforms. The classic veteran in the media nowadays is portrayed with some vestige of his uniform on, usually a military cap or hat. I have a newspaper clipping in front of me now of an old fellow, my age, who is complaining because D-Day, when the allies invaded German-held France, is not being celebrated. He is wearing a civilian suit but has on a military cap, bedecked with medals. He is offering a military salute for the photographer. It is obvious he remembers his military service as a time of glory. I should mention that he was an airplane pilot during the war. Also it is worth noting that most veterans of that war wouldn't be able to put on their uniforms because of their elder age spread.

Anyway, one doesn't need to have been one of the "flyboys" to have the same attitude. I remember that my father had kept his infantry uniform until his dying day. Someone, either he or my mother, would take it out of the trunk to show me. I am sure he couldn't have worn it because of his elder spread. In contrast, I had no interest in my uniform or in the couple of medals I had been given. I think I used the garments as work clothes and tossed them out when they became worn.

So what about "old war buddies?" Did I not have them? No, for a couple of reasons. First, the war-makers in Washington had changed that aspect of troop policy between WW I and WW II. It was decided that replaceability was more useful to the system than camaraderie. So each military unit was made up of people from all parts of the country. And after our discharges, we never saw one another again unless we made some special effort. We couldn't meet one another on the street in our home town, as my father did with his former comrades. And among other changes, many of us left our home towns to live elsewhere. If I made any promises to keep in contact, I never kept them, except one. And that turned out to be a fiasco since our respective lives had already diverged greatly. I had known a New Yorker in the service by the name of John Pulie (pseudonym). He had given me his address and phone number. So when I went to Columbia University graduate school, I looked him up. By that time I had discovered anthropology and was interested in new experiences and ideas. So while with Pulie and my wife on the streets of New York, I saw a vendor of Communist newspapers. I said, "I'd like to get one."

Pulie, "What for?"

"Just to see what they say." This when Communist was the dirtiest word in American English.

"But you wouldn't want anyone to see you buying it, would you? What would they think?"

"Frankly I wouldn't care." Even so, out of deference to him, I did not get a paper. But that exchange was enough. We never saw him again, which is probably just as well. We had little in common except the brief war experience, which I was rapidly putting behind me.

I do remember some of my war buddies, however. There were my two friends from Iowa who went through radio school with me. McGill was a freckled, jovial son of an Iowa farmer and his friend was a Winnebago Indian. I have long forgotten his real name but we always called him "chief." This was long before I had studied anthropology, when I learned who the Winnebago and other Indians were. And I had learned that during the White takeover, all Indian warfare was outlawed while conditions were set up to take the young men into our wars. The term "chief" was a universally used friendly word for an Indian at that time. Then there was Sonny Stubbs, a fellow who had one interest only, women. He spent all his time cultivating the ladies of the local French brothel. He would continually haul pilfered army provisions to them in an army jeep. Then he would use his credit for sexual service. He always wanted me to use their services also, though I never did because of shyness and moral scruples. I had proposed before leaving for overseas.

However, my scruples did not prevent me from having the wildest sexual adventure of my life, perhaps the most memorable occurrence during my military stint. I was taking a walk in the French countryside one day with a tall, thin fellow, whose name I have forgotten. He met a French girl and flirted with her. The three of us continued along the stream together. My friend kept trying to talk to her in broken French. Being shy, I stayed on the outside, planning to return to camp alone. However, after a while he pulled me to one side and whispered that she really was interested in me and that he thought it best for him to return to camp. And so he did, while I continued on with her, trying to communicate with my limited vocabulary. I remember I pointed a lot. And though my moral scruples were bothering me, I became more and more excited. And to make a long story short, after thirty minutes or so, we sat down, then lay down, then had intercourse.

I was torn up by a combination of excitement and scrupulousness, and this, my first experience at sexual intercourse, had to have been very clumsy. I remember she kept asking me to kiss her but I would not, presumably because I thought I would be violating my engagement vow too much. Even so, she passed out, I presume from her orgasm. I was frightened, thinking I had hurt her, knowing nothing about female excitation. I remember patting her cheeks in an effort to get her to open her eyes, but with no success.

Anyway, I finally came to believe that I had hurt her seriously, so I got up and began walking away. Then I broke into a run. I refused to look back and did not stop running until I was out of sight. I really felt dreadful about the whole affair. But then several days later I received a note from her. I presume I had given her my military address, or at least my name. Her letter was from Paris, which had then been liberated. She invited me to visit her. For bait she offered me a German "pistolet." At that time we G.I.'s were collecting war souvenirs to take home. I felt so guilty, however, both for having done this while being engaged, and for having abandoned her, that I never went. If I had, my life would probably have been very different.

According to the vividness of my remembrance, this must have been one of the most exciting events in my military career, a combination of youthful passion and ineptness. And were there no truly military events, including combat? Not really, but then I had spent most of my time in training, and for the year I was in the military theater, I was in the branch that only facilitated the violence of other branches. Our airplanes had no armaments, though we did carry side-arms while on duty. I don't know exactly why, though presumably we could have used them if we had been shot down. Anyway, one of our jobs was to carry paratroopers or pull gliders to drop zones. This was always at night, and as soon as we got rid of the troops, we returned to our home base. Another job was to carry jerry cans of gasoline to tank units at the front, and wounded soldiers back. This was late in the war when the Germans no longer had airplanes in the air.

I was in one accident which could have killed me, but fortunately I did the right thing and survived to become a ruminative veteran. I can't really call it heroic, I was just trying to save my skin. On one of the take-offs, when the plane had a full load of gasoline in jerry cans, the pilot stalled the aircraft. It hit the ground, spun wildly, then came to a stop. I was in the radio booth which was behind the cockpit. My seat belt was attached, so despite the wild careening, I got only a scratch. When it came to a stop, I could hear the fire crackling. I thought there would be an explosion quickly. I remember thinking of this clearly, though it is possible that after describing the occurrence many times in the last 50 years that I am merely repeating my own words.

Anyway, in my remembrance, after unhooking my seat belt, I looked forward and saw the rest of the crew crowding under the escape hatch above the cockpit. I would have to wait for everyone else, which included nurses. So I decided that despite the gasoline in the cargo compartment, I would be better off going that way. I ran across the tops of the jerry cans to one of the emergency doors, turned the handle, and gave it a mighty kick. It flew off and I jumped out running. The cans began igniting behind me, flying through the craft's aluminum skin like rockets. I ran until I felt safe. Someone came up and offered me a cigarette. They did that during the war. And since I had started smoking in the service, I took the weed and smoked while watching the fireworks. The airplane was totally destroyed. Someone took pictures, of which I was given a set. And that is the most heroic thing I did in the war. Probably my father and Hans' father did more -- though for what good purpose, I know not.

    I suspect that most soldiers had careers more like mine than like those of my father or Hans' father. There was a long dreary period of training, then months and years mostly of waiting, and perhaps with one or a few periods of violent combat. Most of it was not only not exciting, or made up of heroic occurrences, but it was boring. Of course there were a few pilots, supported by multitude of mechanics and other people, and the killer elites (marines and special forces), but only in an all-out war like WW I and II were they involved extensively in bloody fighting. The news I hear nowadays from nearby Marine Camp Pendleton is that if one or two marines are killed in a "police action" against a ragtag third world country, it is enough to cause consternation among the troops and their wives. The most important issue seems to be that they should not be separated long from their families, and especially during Xmas or other holidays. From what I read in the paper, these are not the gung ho types who captured Iwo Jima or Tarawa. Those fellows, like me, my father and Hans hardly expected to be back at Xmas.

    So the question is why we hear so much in the media about the heroic actions of our veterans when in actual fact most spent boring, uneventful careers. And the first answer is that violence and heroics are more exciting and thus more newsworthy. It is the same reason newspapers concentrate on crime and other kinds of violence. It is well known that most lives are made up of commonplace repetitions of humdrum events. There is not much excitement. Which is of course why we seek excitement in the movies, television and reading material. Let's face it, romance and violence are the two chief ingredients of the popular media. How many people live in a state of true romance most of their lives? Which explains the popularity of

Harlequin novels, Madison County and the soaps. And how many men eliminate bad guys with "magnum force," which explains Eastwood, Rambo and Bond?

The second factor is that we live in an era of mega-wars which need conscripts or recruits for conflagrations about which they know little except who the enemy is. Even our so-called "police actions" involve thousands or millions of men. Communities have always brain washed (indoctrinated) their members. But in the simpler ones the individual knew more about why he was doing what, including raiding an enemy. Generally, they were displacing others. Before the highly organized mass army came into existence some 5,000 years ago, when tribal raiding was the norm, fighting probably had more excitement in it. And at least the Sioux who was raiding a Cheyenne encampment understood more why he was doing it. But modern wars are worked up by war departments, elite in positions of power who decide when and whom to fight. However, they need fighters who will not question their decisions. Moreover, we live in an era of opinion molding which the elite have primary access to. So it is almost inevitable that this elite will do everything it can to mold public opinion. And this will build on the brain washing that has been going on since birth. All the resources available are used. Thus we have Jimmy Stewart, Bob Hope, multitudes of nubile maidens, and hordes of other entertainers. And it is perhaps inevitable that deception will be used to mold public opinion. Horrendous acts by Hun savages to Belgian women and children, and Bay of Tonkin incidents are natural. None of this is original. It was all spelled out by Machiavelli who of course was villified through history for his honesty.

And of course such brainwashing works best on the young and inexperienced. None of my three "heroes" questioned it when their time in the military came. My father, probably typical, never changed his opinion and Hans' father never had a chance. I am the only one who did, but I suspect that the odds are far greater than one in three. I would suspect it to be closer to one in thousands, of those who became thinkers and looked at the social system critically. And for that I am grateful, not that I believe it will change anything. After all, Machiavelli did not, why should I? But at least I got to tell my father's story and the little that I know of Han's father, plus my wild sexual adventure in France.

24

On Getting Older

I saw a bumper sticker recently which proclaimed " Screw the Golden Years." I was intrigued because after thinking about it, I realized I had reached that stage of life, though with a different outlook. It seemed worth putting down on paper.

So I will begin by carping. I find the term "golden years" as unattractive as the other one, "senior citizen. I suspect that hucksters in the business of selling something were involved in devising both. I don't like "golden years" because it emphasizes that less than noble American proclivity, greed for material wealth. In our later years we will have gold or be like gold, even though the metal has little value for humanity except for gathering, trading, and storing it. The term tells us nothing about the other qualities of life -- excitement, curiosity, lust for learning, lust for using the body. creativity, and connectedness (with the world and others about us). And "senior citizen" tells us no more than that without losing our membership in the body politic (our nation), we survived many decades.

There are less complimentary terms for older persons in American society. Behind their backs or from one car to another, older men are not infrequently called "old fart," "geezer," or "pops" by younger men. This particularly happens when the older man gets in the way of the younger. Presumably, "old fart" implies that the older fellow is smelly or less than clean. My dictionary says "geezer" means "eccentric old man," which is not all that helpful. "Pops" "and "gramps" seem to be relationship terms, though my impression is that they are generally disrespectful. The formal honorific for the progenitor is father, a little less formal but still friendly is dad, while pop brings the man down to the level of the child, with a humorous implication. One doesn't use the term, pop, to a person from whom one asks permission to take a seat. A term which doesn't indicate deep respect but does imply strength is "old man." One doesn't mess around with the "old man." And finally "pops" takes away practically all respect, but keeps the meaning of father. I find "pops" a very natural term to have evolved in America since relationship roles have become so unimportant. One doesn't ask permission of the older person in the family to sit down or smoke a cigarette, as happens in traditional patriarchal societies. The typical white middle class attitude is to expect dad to be a kind of buddy with money and things, but certainly not to be a disciplinarian or rule maker.

After all these terms with negative connotations, it would be unfair of me if I were not to discuss the one term for older men in America which is respectful, or at least not disrespectful. This is "sir" which I first noticed directed toward me when I was well up in my fifties. "Can I help you sir?" the kid in the supermarket would say, or "Sorry sir," by the young person who bumped into me. "Sir" is of course an honorific used towards those with position, money or power, even when they are not very old. And though it may be used spontaneously sometimes, it is also certainly taught to clerks and other job-holders who are expected to keep members of the public in good spirits. I presume the term, madam, is also taught to be used for the same purpose. Young people in fast food restaurants invariably address me as "sir." Also the traffic policemen that have stopped me have been very punctilious in this regard.

I remember a particular exchange I had with a traffic cop a few years ago in which I was "sir-red" to death. I had been flagged over on a Los Angeles freeway, and after the customary procedures had been gone through, and he had taken out his ticket pad, I asked the man why I had been stopped. He said, "You were traveling too slow, sir."

"Too slow? I was going about 55."

"You should have moved to another lane, sir. You were in the fast lane. You should have moved to the right."

"I couldn't. There were cars alongside me." I am not sure there had never been an opportunity to shift lanes, but there had been a lot of traffic.

He continued writing. I felt exasperated. It seemed too much to get a ticket for driving too slow. I had not had one for 15-20 years. I said, " Don't you ever give warnings? I'm a very safe driver."

He stopped writing and looked at me. "It's up to the discretion of the officer, sir."

My frustration grew, though I tried not to become belligerent. I said, "I know it is at the officer's discretion. But when does his discretion make him give a warning instead of a ticket?"

This time, without looking up, he said, "When it seems appropriate -- sir." I felt it had become hopeless to talk about this any further so I stepped over and leaned against the car. He continued until the ticket was filled out, had me sign it, and gave me my copy. As I took it, he said, "You know you're a real gentleman, sir." I felt like laughing or crying, but said evenly, "That's not much help though, is it? I still get the ticket. What if I weren't such a gentleman?"

He said, "Oh, you don't know what kind of people we have to deal with, sir. Why I once gave a ticket to a guy who got so mad he started picking up gravel and throwing it at his car. I just stepped to one side and let him hit his car. After all, he was damaging his own property. I gave him the ticket after he was finished."

I just couldn't take anymore. I laughed. "It seems that people you stop get tickets whatever, no?"

He said, "As I said, sir, it's at the discretion of the officer."

Whereupon we got in our vehicles and went our respective ways. I will admit that police officers have a special problem, which is that while they are supposed to be serving the public, they are generally punishing the individuals they accost. I suppose the theory is that verbal politeness makes up somewhat for the aggressiveness of their actions.

I knew I had moved into the world of older persons when I stopped in a McDonald's a few years ago. I was on the way home to my avocado farm in San Diego County after a long day teaching class in Los Angeles and I needed a caffeine fix. I ordered coffee from a young girl who was pertly dressed in the standard uniform. She asked, "Are you a senior citizen, sir?"

The question took me aback for a moment, since I had evidently been pushing this idea to one side. I had noticed for many years that on the price list of theater marquees and other ticket windows there had been a rate differential for different kinds of persons. There was a reduced price for those over 60, students, members of the armed services, the disabled, and some other categories. Anyway, I had ignored these lists and paid the full price, presumably to keep me (and others?) from recognizing my advancing years. Another incentive to pay the regular price was that I usually came to the theater with younger persons, ordinarily women. They were invariably 10-30 years younger than me. At first it seemed embarrassing to ask for one regular price ticket and one for an older person. I loosened up some when I realized that the women who accompanied me really didn't care about the difference in our ages as much as other characteristics. I remember in fact of making a joke about our age difference a couple of times. When getting tickets for myself and a younger female companion, I asked for one ticket for a student, which she was, and one for a professor, which I was.

Anyway, there I was being questioned about my age, while trying to get coffee. I knew I had to admit my age someday, so I decided why not do it when I was alone in a place where I would probably not come again? I said, "What age makes me a senior citizen?"

"Over 60, sir."

I said, "Okay, that I am. So how about the free coffee?"

She filled a styrofoam cup with the watered-down potion favored in American fast food joints and I took it to the little stand where one got sugar, artificial milk and plastic stirrers.

As I stirred the light brown liquid, I wondered who had decided on passing out free coffee to oldsters. Presumably there was a sales gimmick involved. MacDonalds was a very successful franchise, and there were undoubtedly few practices which were not expected to pay off at the cash register. Anyway, this particular practice for this particular McDonald's did not seem appropriate to me. Here was a kiddo who worked for $4-5 per hour, offering a free coffee to someone who when he worked made 5-10 times that and did not need to work at all. I was semi-retired, teaching one-third time, primarily for intellectual stimulation.

And though there is undoubtedly a fair number of older persons who live somewhere around the poverty level as defined by government bureaucrats, they must mostly be in the big cities and usually in the poorer sections. However, even in the cities there are fair numbers of older persons who are not so impoverished. And when one reaches a place halfway between Los Angeles and San Diego, one has traveled far from the inner city. The 50 miles circumference around the area of that particular McDonald's, in Corona, has mostly middle class homes or condos. They are not the residences of down-and-outers. Many residents are retired people who have had enough money to buy outside the smog zone.

As one moves farther away from the city, one gets away from the tract developments and finds oldsters better off. In my neighborhood of northern San Diego County, most are owners of houses and property who grow avocados as a sideline but subsist primarily on their retirement and investment checks. One of my neighbors told me he and his wife were "double dippers," a couple who had two pensions and two social security checks per month.

A few work at full-time jobs. Some others work hard at hobbies, and some even make businesses out of their fun work. One retired ex-Angeleno is one of the foremost tropical fruit growers of southern California. Another from Los Angeles breeds all kinds of exotic animals. His latest have been buffalo, longhorn cattle and emu. He has dozens of smaller creatures, plus equipment to raise every kind of fish, fowl and mammal. I have another neighbor who is a retired naval officer and who spends his time in his new home on a hillside, out in his power boat off Coronado Island or vacationing in his mobile home. His hobby is construction and fixing things. Most days he can be seen peeling up and down the road in his pick-up, hauling things for the next job. Apart from repairing and installing several small items for me, he built a fine greenhouse. He seemed to enjoy every measurement he made and every nailing he hammered in. He is a friendly fellow and will talk, but not so much as to interrupt his work-play.

Then for those who like their retirement places pre-arranged, there are the golf communities. Up and down every road hereabouts there is a golf course with attached developments. The older golfer has all facilities for his recreation outside his door throughout the year. Most such developments have bars, restaurants and shops also. The granddaddy of these communities is the Lawrence Welk Resort, which has all the above, plus a theater, legal service, dental service and more. The oldsters in these communities do not need free cups of coffee. This is not to say that they are rich people, but most do not have to stint, and short of a national disaster and the collapse of the monetary system, most will have few financial difficulties the rest of their lives.

Some of the hucksters in our society have recognized the existence of well-off oldsters. They have given them a name, the affluent seniors, and directed sales strategies toward them. And if we can believe anybody about sales, we should believe the hucksters, because their success depends on their accuracy. Salespeople who fail to analyze their clients well do not sell their goods.

I have wondered why we hear so much about oldsters who eat dog food and live in vermin infested apartments, while we hear relatively little about those who live comfortable lives. The best I can make it out is that our news media thrives on adversity. The producers of news programs and specials like to show us misery and disaster, and evidently most of us like to see it. We dote on programs showing people in trouble.

Our newscasters go wild about hurricanes and other storms. A few years ago the hurricane, Hugo, passed through the Caribbean and left its mark on several islands. In their reports the national newscasters I saw almost frothed at the mouth with florid adjectives. The storm kept "wreaking havoc' and the winds "wailed like banshees" while galvanized metal roofing sheets became "lethal weapons as they sliced through the air like knives." I can imagine the newspeople discussing the event before going on the air. The anchor man says, "What's the lead story tonight, fellows?"

"Oh, it's good. The hurricane went straight through Puerto Rico. For a while there it seemed like it would miss San Juan. But it veered back and turned the streets into oceans of water. There are great shots of flying and floating debris."

"Sounds like we got one story anyway. Any human interest?" So we dote on violence on TV. I suspect the reason we like cop shows so much is because there is so much fighting and shooting and accidents. Even in our dramatics love and mayhem are the main themes. So perhaps we should approach the question about suffering oldsters from the other direction. It should be, "Who wants to watch people living contented lives?"

I shouldn't take credit for this observation since it has been around a long time. I remember in my early writing classes that the professor would declaim the necessity of creating a problem for the main characters for dramatic effect. My only contribution is that it helps explain why we see so much more about the dismal oldster, living on the margin of respectability, safety and security. I should also point out that bureaucrats and professional do-gooders make their living by running programs to help the less fortunate. That is their way of keeping in the better-off class themselves. So they make noise about poor oldsters, which in turn is amplified in the media by persons keeping in the better-off class by showing endless adversity.

Anyway, by some quirk of fate, I ended up in my later years more comfortable financially than at any time in my previous life. It was not a consequence of deliberate effort. I never aimed to be a millionaire and probably never will be. And yet from the penurious days of the Depression, to the comfortable 1980's and 90's, I tried to take the best measures I knew for maximizing my wherewithal. Becoming a rich man had never been a primary goal, and yet letting what came in to dissipate mindlessly bothered me. As soon as I realized how much one used up in credit buying, I switched to cash and have continued ever since. Several hucksters have been made uncomfortable by this custom of mine.

I remember a car dealer once in the 50's who, after settling a price with me for a used car, asked me to come into his office to settle the purchase. At that time, auto agencies frequently carried the loans for the cars they sold. In any event, this car dealer sat down in the pompous way of car dealers, and after shuffling through papers for a bit, said, "And what kind of terms would you like?

"What do you mean?" I asked.

"What kind of loan do you want?"

As I recall, the total cost of the car was around $400, but in those days that was a sizeable figure. I said, "I'll just pay cash."

"What do you mean?" he asked.

"I mean I'll just pay you the money." I added, "I'll go down to the bank and get it." In those days our family finances were so simple, we didn't even have a checking account.

He bridled. "I didn't offer you that price if you were going to pay cash."

I replied, "You didn't ask me how I was going to pay for it."

"But almost everyone who buys a car here buys it on time."

"Are you saying there's a different price if the car is paid for outright?" I hadn't really thought about it before, but I then realized that this would make a real difference to the agency. They calculated on the additional 15-20% interest they would get by selling the car on time. Anyway, the man did not insist and did give me the car for the agreed price paid in cash. Now that I look back, I remember several such incidents, which indicate caution in financing.

I am aware that most economic thinking in this country is based on credit buying, including paying sizable interest fees during most of one's life. Until quite recently taxpayers were allowed to deduct such interest fees from their income tax. But without any expertise in economic affairs, I still fail to understand how paying this additional fee helps either the individual or the economy in general, though I can see how it helps the loan agency. Furthermore, there has evidently been some change in bureaucratic thinking, since the deduction for interest paid has been taken away.

And although I don't think it has added up to much in dollars, I know I've adjusted my buying to the cost of purchases through the years. I still can't help checking the price of an item in a store and refusing to buy if it is high, no matter that I can afford it. I know this is a direct heritage from my upbringing during the Depression, when my mother had to use my father's meager salary with care. She managed and I was indelibly marked. And though I certainly can afford to now, I still find it hard to buy fruits and vegetables out of season. Bananas usually sell for 15-25 cents per pound, but this year their price is more often in the thirties and forties. I can't bring myself to buy them. Fortunately, I have a fair amount of my own bananas now growing on the farm. Jessica, who knows the Depression only as past history, has no such buying problem.

And did I invest as some money accumulated? After all, that is the classic American way, even if by the old fashioned method of using one's own money. The super investors become millionaires through the use of other people's money. Much to the damage of the productivity of the United States, I'm afraid they have developed a system called the leveraged buyout in which for a pittance of their own, and with quite a bit borrowed from banks, the supers can buy out giant corporations, to merge them for further profit. Donald Trump has replaced Henry Ford. Wealth thus comes from manipulation of money, not production.

But to get back to my case, there was no super-investing and hardly any general investing. The main problem was that I found investment procedures boring and refused to read financial reports. I did have a brief period when I worked through an investment agency but I found that less than satisfactory. The stock broker would call me and suggest that I buy or sell so many hundred stocks because of such and such a development. He would then send me the printed brochures. I'm afraid I got through very few before I decided to go on the basis of his opinion. Thus I really didn't know enough about the companies to understand why I was buying this or selling that.

I'm not sure now that many high powered types know much more. I believe many of the super investor/money manipulators know little about the companies they buy and sell. And of course the reason so many banks have gone on the rocks in recent times is because their investment officers frequently knew little about what their loans were going into, nationally and internationally.

However, my lack of information bothered me even though I wasn't willing to put enough effort into getting it. Also I knew that stock brokers made money every time there was a transaction; so why wouldn't they want me to buy or sell? Anyway after a couple of years I got out of that business and put the money into real estate.

I also had a brief fling with the gold market once, again with no notable success. A professional colleague (college professor) convinced me to buy some on the basis of a popular guru who was making a lot of money selling books about investing. Fortunately, I did not buy a tremendous amount and I'm afraid I had about as little interest in its ups-and-downs as I did in common stocks. After the value of the gold coins had dropped a little, I decided to get out of this form of speculation. I had a necklace made of them for Jessica. She had even less interest in investment procedures than I did, but she had considerable interest in jewelry. It seemed a fair solution. The gold did remain in the family as long as it lasted.

The one kind of investment I did take seriously through my lifetime was real estate. This always did seem better to me, since a house and its environs was something real which I could look at and judge without being bored out of my skull. Furthermore, I could live in it, enjoy it while still investing. Also, there was a considerable income tax advantage in home mortgages, plus when I got to California quite an increase in value. But I still think my major decision always was for the pleasure of using the place. The investment potential was simply an added benefit. Thus, the house on top of a hill, the country place from where I am now writing, was undoubtedly a good investment property but I bought it to live in and for therapy, purposes it amply served.

Anyway, my residences have done me well financially. And being in southern California certainly helped. Although there are other urban centers where houses have appreciated a great deal, I know of none that increased any more than in southern California.

So here I am, financially comfortable, though far from rich. Is that all that is needed for a contented life in the later years? I am sure it is not. Being forced to stint and forego much must put some kind of burden on an oldster. But what is acceptable economizing or degrading poverty I hardly know. Certainly my parents economized and did without many things most of their lives, as did most Americans of their generation. But I never saw this as one of my parents' major problems. And I'm sure that the primary satisfactions of my life did not come from affluence. There were as many good times during the lean years, even though it has become comfortable to get further away from poverty. Now money matters can simply be taken for granted.

So what is it that makes the good life, especially as one gets into the later years? I must say that I never worked out a plan in advance or even thought about the problem consciously. Like most others, I thought of the good life and tried to get it, but normally only for the period I was then living in. I made most life decisions because I thought they would make life more interesting. Unlike most of my students, I did not go to college because I thought it would help me get a better job. I really liked to learn all those new things in books and classes. Perhaps this is why I ended up in anthropology, a field which is known for being interesting but not noteworthy for helping one get well-paid jobs. I did not even think in terms of the future in things like using money judiciously. It was almost a reflex action to avoid overpaying, I suspect a direct heritage from my parsimonious mother and a consequence of living during the Depression.

So what else is needed for a ripe old age? And although I can hardly take credit for the idea, I'm sure good health is near the top of the list, higher than financial security. Thus, it is little wonder that the medico is ranked so high in all cultures. Anthropologists long ago discovered that the one person with special status and privilege was the medicine man. Even in the simplest societies the curer always stood out. And of course this situation only became more formalized as society became more complex. The medico is still up there with the controllers of power, the politicos, military and cleric in the receipt of prestige and rewards.

It is not difficult to understand. One needs to feel good to do anything, from the least taxing mental exercise to the most strenuous work. If we are sick, we cannot even use our money adequately.

This creates a special problem for the olding. Their bodies have been long used. Many parts are wearing out or have already done so. And so what can they do? As I see it, there are two main ways to maintain health. The most obvious one is to use specialized help from the medical profession, primarily to correct body problems as they come up. Fortunately, our medicos have become quite good in this respect in my lifetime.

After my most recent surgical experience (total replacement of arthritic hips), I thought a lot about this and realized that I would be crippled or dead if I had not received the medical help I did in my lifetime. Nothing particular happened surgically when I was a child except the loss of my foreskin and tonsils, operations which were done almost automatically in those days. As a young man I had two hospital stays, one for pneumonia and one for mononucleosis. I presume my condition was fairly serious in both instances. I was in the military during the first and a college student during the second. I think either could have killed me if I had not been treated. I must have been in good health for about fifteen years after that, though I did have frequent pain in my lower back. This turned into a herniated disc, after which I couldn't stand erect. In a peasant or tribal culture I would have been crippled the rest of my life. In the U.S., however, after a week or so in traction in a hospital, and a couple of years of therapy, I regained normal posture and strength.

Not many years after, I had a ruptured appendix which was

surgically removed very quickly. If untreated, this could certainly have killed me. Another 14-15 years of body ease and whammo out of nowhere -- a stroke. I had never known I had high blood pressure, though I should have guessed it, since my father, his father, and several of my father's brothers died from it. Anyway, this was a serious one and shook the bejesus out of me. And again fortunately, the medical profession took over. I have no doubt that I would be either dead or paralyzed if they hadn't worked me over those 8-10 hours in the emergency room.

That was my last life- threatening ailment, though hardly the last potential crippling one. A few years ago my joints began to go out, and after three years, I went back under the knife, to have my hips replaced. I would now be hobbling about, a bent old man, without that intervention.

I am leaving out surgery which simply made life more comfortable. But while I'm at it, I should mention that at least three other crippling ailments were corrected, broken bones in the finger, femur and elbow. And although we don't usually think of dental conditions as so serious, I have had considerable trouble with my teeth, and extensive treatment. I now have dentures, which is an ideal solution. I can eat anything, need to do a cleaning only once a day, need do no flossing, have total security, and no aches or pains. Moreover, I practically never have to see a dentist. I should have had it done years before instead of the multitudinous methods of "saving my teeth" advocated by a variety of dentists. I hate to think of how much I paid to avoid going to dentures. These days I see myself as more or less a bionic man. The techniques for replacing parts and buffering undesirable conditions have progressed almost more rapidly than my body has deteriorated. I have no doubt that the actions by medicos have kept me from being a cripple or dead several times. But not being dead or crippled does not mean being healthy.

To be healthy, and especially in the later years, must be a condition which enables the individual to be a participant in some significant events, to be creatively active. This requires body energy and a wish to do something significant. To mope around is not health. And correcting life accidents alone will not make the difference. What will are our self-chosen personal activities, of which we get least help from the medicos. They are too busy with the accidents.

And however good they have become in administering drugs, cutting out, repairing or replacing body parts, our medicos have been far less successful in helping us use our bodies when we are on our own. And unfortunately these activities are the main ones which will make the difference between health and illness. Furthermore, their consequences will accumulate as we grow older.

What are these activities? I think that they are mainly what we eat, what exercises we do, what other habits we have, and how we think.

I feel sure that we are what we eat, the more so as we get older. I give the medicos credit for figuring this out, even though they never really learned much about how to get us to eat the right things. Probably the best they have been able to do has been to threaten us with recurrence of some illness if we don't change. But frequently when they get to that stage, it is too late. Eating primarily for promoting health is a real problem, especially in a culture of affluence. Probably the greatest negative in this regard is that we can afford to eat so much. Furthermore, our nutrition specialists have failed to recognize the significance of the fact that we don't eat primarily for health. They have emphasized which proportions of which chemical components we should take in: proteins, carbohydrates, fats, minerals, vitamins, etc. But these are not what we eat. We eat substances that satisfy our sensory cravings, the concentrated body flavor of sliced and ground muscle (burgers, steaks), the viscous thickening of fat (french fries, donuts), and the melting sweetness of sugar (milk chocolate, ice cream). We eat primarily for pleasure, and unless our diet is critical to survival or our social status, we only consider the calories or amount of fat to protein after we have decided whether or not the item is yummy-yummy good or icky. And thus we have the problem of eating too much while the cultures of deprivation have the problem of not having enough. Even the poorest of us can afford all the fats, carbohydrates and proteins we want.

I have long been amazed by the appearance on the media of those who protest welfare cutbacks. A high proportion are fat. Whatever else they suffer from, caloric intake is not one. Hamburger, french fries, refined bread, dairy products and sweet rolls are available to all.

It is true that we have a couple of illnesses, bulimia and anorexia, that are the consequence of eating too little. However, compared to those who are overweight, those who starve themselves to fulfill social and psychological needs are in small numbers. The weight illness of the affluent society is heaviness, as the weight illness of the poor society is thinness.

And how does this problem affect the olding? First is that the older body can less afford to carry extra weight. Like a society, a body is a living machine with inter-related parts. This means that any part out of adjustment is likely to affect other parts. When muscles, including those of the lungs and heart, have been used for 60-70 years, they are less able to carry their body up a long flight of stairs or even on a long walk. And if extra weight has to be carried along, the strain is greater.

Moreover, it is more likely that older persons are going to be heavier. There is not as much pay-off socially for oldsters to be slim. They are not usually in the dating game, so it isn't necessary to impress others with their sexual attributes, probably the main reason for staying slim. Moreover, older persons have the disadvantage of continued habit. They tend to continue eating the same amount they have been eating through their adult lives. And lastly they are less active than younger persons, so their body burns up less.

Somehow older persons have to learn to cut back on their food. But since so much dieting goes on in modern society, this would hardly seem to be a problem. But of course it is well known that the majority of dietary regimes merely make money for the organizers. My mother was on or off a diet as far back as I can remember, and when she died in a retirement home she was well overweight. Most of us have had friends or relatives who were on one diet after another. It's a veritable roller-coaster. Unfortunately, the truth is that it isn't easy to break a pleasurable habit. And eating is probably the most pleasant there is.

There are undoubtedly a few fortunates whose bodies adjust to the advancing years without special regimes. But everything being equal, it is likely that if they don't cut back, most older people will gain weight. I think I generally lucked out in this regard. Throughout my adult life I ate standard American fare with gusto and remained at about the same weight. I was normally active for a white collar type (college professor) until I took up bicycling again in my late forties. But then two experiences changed my eating habits. Probably the most important was the first, I got divorced. But unlike many divorced males, I liked cooking as I had ever since boyhood. Anyway, I took to making a lot of innovative dishes. Also at this time, my early fifties, I started dating again. I became quite conscious of my appearance.

The other change was that after a few years I gave up eating all meat but sea food. It wasn't for dietary reasons but because of my feelings about all the animals that were being killed so I could eat their parts. While I was married to Jessica, I gave up fish also. I must put on record, however, that I do not believe eating meat in itself is fattening. During most of man's prehistory, almost five million years, he was a hunter while his woman was a gatherer of vegetal products. And though we have no direct information about prehistoric man's health, we do know that the diet of hunters that were still managing in the 20th century seemed to be more than adequate for their health. This even includes the Eskimo, one of the heaviest meat-eating peoples described by anthropologists. However, all primitive hunters lived on the meat of wild animals, which is probably far more healthful than the tender, fat-striated meat of industrial farming peoples.

Anyway, I became more health conscious about the same time I was changing my other eating habits. Among other things, I broke the much more addictive habit of smoking, taking all my pipes into the garage one day and breaking them up with a hammer. This stopped my pipe smoking, though I still continued to buy or snitch cigarillos and cigarettes. It took another 5-6 years before I stopped smoking completely. By comparison, giving up meat was easy. And so, sometime before I got into my late 50's, I cut back on the amount I ate, out of health consciousness and to try to keep slim for the dating game. I am now a little heavier than I was in my early 20's but I think I'm in the "normal" range for my age.

I have long thought about weight control, and particularly diet plans, and like many others, I realized they usually do not work over the long haul. So for better or worse, I will offer my own system. It's not complicated and doesn't take away the pleasure of eating. Moreover, my plan is free. I won't call it a diet. It is so simple, it couldn't be sold. Periodically I get on a scales and notice that I am 5-10 pounds over my lifetime weight. So I merely cut down on everything I'm eating. If I am then in the habit of eating one or two pieces of toast with an omelette, I cut out the toast. And if I have been eating a salad with two helpings of macaroni and cheese, I cut the cheese down to one helping and the salad into one-half or out. For my morning tea, I put a level teaspoon of honey in instead of an overflowing one. Moreover, I quit between meal snacks except for an occasional half-glass of milk or some tasty variety of cracker without cheese or peanut butter.

It is amazing how fast poundage comes off with simple reduction measures like this, about two pounds a week until I get back to a "normal" weight. Moreover, I get to continue eating the goodies, simply fewer of them.

I am not the only one who uses a simple procedure like this. I think many ladies in our society who are concerned that they keep a slim figure do the same, or some variation thereof. And let's face it, females concerned with their attractiveness are the supremely successful weight controllers of our society.

And is that all there is to it, to cut down one's eating when the years accumulate? No, if good health is the goal, there's still the quality of what we eat. We of the "supermarket civilization" have a real problem here. On the one hand, we undoubtedly have the biggest variety of foodstuffs available of any society there has ever been. About the only things not available are those which our cultural taboos keep us from eating. For instance, about every kind of meat there is comes to our markets except the forbidden ones, like dogs, cats, horses, insects and most reptiles, many of which are relished in other cultures. There are fewer taboos among plants, though because of unfamiliarity, many good ones are not to be found in our produce departments. Most Americans just aren't used to fixing and eating breadfruit, banana blossoms and quinoa seeds. But still when one looks through the canned and frozen food sections of a supermarket, one quickly becomes aware of a very great variety.

To consume many different kinds of food is undoubtedly a healthy thing to do, the reason being that what some lack is made up by what others contain. This is especially true if more than enough is consumed. The body can then select those elements it needs.

It is generally believed by anthropologists that the diet of primitive hunters and gatherers was nutritious. Because it was not easy to get enough food with this kind of technology, hunters and gatherers tended to use most that was edible. We have particularly full accounts of primitive peoples who lived in arid or semi-arid zones, such as Southwest Africa, Australia, and the Great Basin of the American West. Such people used all kinds of wild plants which would be passed up by cultivators. They also ate every kind of wild life which could be caught, including rodents, reptiles and insects. The Indians of the American Great Basin were looked down upon by whites, who called them "digger Indians." However this grubbing, digging out things, using most everything found, was undoubtedly good for them. Their problem, like that of other gatherers, was periodic scarcity of food.

One of the nutrition problems which emerged when man became a cultivator was that in various parts of the world he specialized in a few plants and animals and developed preferences for them. Thus all cultures got food biases, depending on what the ancestors had developed. The Italian-American eats a lot of pasta with rich sauces and cheese, the AngloAmerican wants large pieces of roast meat with well-cooked vegetables, plus potatoes, bread and dessert, while the Chinese-American goes for rice or noodles, stir-fried vegetables and a little meat or fish. None eat anywhere near the variety that is available. The only people who approach being encyclopedic in their appetites are the Chinese, but they have some taboos also. Moreover, as with other habits, we cut down on the variety as we get older. By our 60's most of us have become accustomed to a fairly limited repertory. Just as we resist learning new languages in our later years, we also resist trying new foods. Probably the most nutritious diet we have in our lifetimes is in our early adulthood because we are then most adventuresome. We have got through the limited variety of foods of childhood and youth, when we tended to latch on to those few items offered first, and have not yet got into the fixity of the later years.

So for a better diet, the older person has somehow got to try to break out of this pattern and eat a variety. And though it is okay to eat food from prepared lists of nutritionists or other specialists, I think it is usually boring. The alternative, to be open to new foods generally, as well as to most other customs, would be much more preferable.

How to do this? As usual, I turn to my own practices which I don't expect others, most of whom have had little cross-cultural experience, to follow exactly. But perhaps the idea is worth while. I was always tempted by something new, including foods. I suspect this is a part of the anthropological personality. We are usually more interested in new customs than the ones we learned as members of our own society. In fact, when I sometimes used to tell others that I felt I was a man of all cultures, Jessica would catch me up by saying, "Except his own." She was partly right in that I, like others in my profession, drifted into study of the way of life of "others" because we were not well adjusted in our own. Anyway, in eating as in other customs, I have long been intrigued by the new. If I were to find breadfruit in the produce section, I'm sure I would buy one and try it. Anyway the oldster will do better if he/she will cut back on amount and eat a wide variety. Mankind has come a long way from his primitive ancestors in the processing of food. Those who live in the supermarket civilizations are no longer constrained to eat only what they have caught or gathered in the last few days. In fact. the great majority do not get their own food at all. They buy it in the store. And though they may be pleased with the idea that they don't have to dig their own wild roots, or run down their own warthog, they do pay a certain price.

Through the inexorable evolution of technology, mankind discovered some basic improvements for the preservation of food. He learned how to use certain substances for keeping bacteria out, he learned how to slow down the process of decomposition. Also he learned how to use bacteria for fermentation (wine, beer) or acidification (some dairy products, vinegar) and medicine (penicillin). By and large, these have been beneficial discoveries for promoting food variety and keeping it over long periods of time. But another discovery, refinement, has been considerably less beneficial. In the last thousand years, man has learned that by taking unwanted parts out of basic foods that they would last longer. So he took the germ and skin from grains and the extra ingredients found naturally in salt and plant juices. In other words, he made concentrates of them. Thus he came up with pure white flour, white rice, salt and white sugar. Moreover, through time the people of urban cultures got used to these pure substances so much that they ate no other kinds. The results were not so wonderful. People ate too much pure salt and got high blood pressure, and too much sugar and got diabetes. But perhaps the one which caused most trouble was refined flour from which mankind got much more dental caries, constipation, obesity, and other ailments. And characteristic of the way things got done in modern, fractionated society, people were then told to increase the fiber in their diet, as if this were a new idea. Their ancestors had been getting much roughage for at least five million years.

Ironically the people who suffered most from refined food were the few surviving primitives. Once their land was taken away by Euroman, and their gathering way of life destroyed, the pre-urban peoples had to go to the new stores for food. Their diets deteriorated rapidly. They became overweight, their teeth decayed, and they got a whole host of diet-related illnesses.

It should come as no surprise that constipation and other digestive problems are more frequent in the older body than in the younger one. The older one has been processing food a long time and has less flexibility to adjust to difficult conditions. The dietary solution is simple, though as usual not so easy to bring about. Basically the older eater ought to cut down on refined food, though how is difficult to say. Many older people are addicted to white bread and sugar, as well as generous helpings of pure salt. I don't know that my father modified any of these practices before he died of multiple strokes. Like many Americans, he was addicted to much salt, white sugar and foods made of white flour and sugar. He put salt on most of his meat and vegetables even before tasting them, he expected desserts for every meal, almost all made with white sugar, and the only kind of bread he ate was sponge white. He did eat all-bran cereal for breakfast to help his constipation problem.

My dietary history is different, mainly from changes made after I was 50. Until then I ate standard American, which included considerable refined salt, sugar, flour and meat. Also at that time I was living as a typical suburbanite householder. All that changed when I divorced and moved to California. I became health conscious at about the same time I began biking and jogging, I quit smoking and cut down on refined food.

What was the cause of my changes? First off perhaps was that I moved to California, where health consciousness has long been popular. There was a lot of publicity in the media about health. Then too I was probably becoming conscious of aging even though the full awareness did not come until my middle sixties. And I suspect the fact that my social life changed so much was important. I divorced and got back into the dating game. I must confess that I did want to look good to the ladies I met. And so, from some kind of combination of these factors, I cut out most of the refined food I had eaten all my previous life. I still used white sugar, but much less, and I switched from white to bran bread. I still continued to use some salt for another ten years. Then I learned by having a stroke that I had high blood pressure. I put the salt cellar away permanently then. So now like the foragers we have displaced in the world, we find it advantageous to go back to some of their habits: to eat a variety, to reduce the amount, and to replace the refined food we have become used to with that containing fiber, the more so the older we get.

There are some other substances we take in considerable amounts, both unrefined and refined, which leave their marks on us. Mankind learned long ago, and in many places, that among the plants he/she (mainly the latter) found in the fields and forests, there were quite a few that did funny things to the body and mind. Parts of many plants could change a person from anything to giving a feeling of well-being to creating a world of fantasy. So she/he selected some out to use for mind altering. Later cultivators began growing some. We call some of them drugs, others we are not so sure of. Anyway, for changing our body-mind state we take a great variety, the only animal that does. They are not foods in the sense that they give us nutrients but they seem to fill a universal function. The main ones, coffee, tea, tobacco and alcohol spread to almost all cultures, despite strong efforts by many religious and government agencies to prevent this from happening. The only clear success to forbid such substances I know of are the Mormons who live in the midst of a drug taking culture besides. The Muslims have been fairly successful in prohibiting alcohol but they are heavily addicted to coffee, tea and tobacco.

So it is not surprising that older persons try to continue taking some of the drugs they learned about in their lifetimes. In some cultures, like the opium producing tribes of Laos, the old men were given relative freedom to indulge since their most active period of life was passed. Young men were supposed to keep away from the pipe, since they still had to be heavy producers.

In the affluent cultures the mind altering drugs also continue to be popular with the aging, in everything from coffee and tea to tobacco and alcohol. In general, the olding do not seem so addicted to hard drugs like heroin and cocaine, probably because these were not widely used when they were growing up. Some people use specific definitions of drugs usually when they are taking one, but not another. My ex-wife, who was a teetotaler but heavy coffee drinker, maintained that alcohol was addicting and debilitating while coffee was not.

Should older persons refrain from all these substances, since their bodies can tolerate less? I'm afraid this will hardly happen, and I'm not sure it should. Some of these substances give people a lot of pleasure throughout their lives and into the later years. So unless an overall proscription with strong social pressure, like that of Islam or Mormonism takes place, men will continue taking drugs, old and young. Overall, mankind has learned that the satisfactions of mind-altering substances far outweigh the negatives. And who is to claim that the oldster has no need to alter his mental state? About all that can be realistically expected of older persons is that they try to reduce the amount in respect to their physical condition. Part of this will be automatic in that the negative reactions of the older body will make less more desirable. The rest will have to be done through determination, on the grounds that more satisfactions will come only to the unaltered or slightly altered mind. Both alcohol and marijuana have been claimed to cut down on sexual desire. And while heavy doses of either may cause such, a moderate use may stimulate sexual activity.

I have taken most standard drugs in my lifetime, as well as a couple of non-standard ones, at least for my socioeconomic status. I am not against taking drugs, though I suspect I have been cautious in taking new ones. I've taken my share of the two standard uppers, coffee and tea, though now I am cutting down on both because they irritate my stomach. I smoked tobacco through my early and middle years but finally kicked the habit. I tried both marijuana and cocaine, and though both were pleasurable, the pleasure was not enough for me to deliberately seek more.

Probably the main drug throughout my life has been alcohol. In my youth I began drinking cheap wine, trying to achieve instant manhood and impress my friends. Drinking was standard behavior in the military. But afterward, when I married a teetotaler, I cut back to please her. I survived my householder's period of life with strain, letting go at the occasional party. But after separation I turned to the bottle seriously. Since then I have probably drunk too much for the best of health, though lately I have cut back. So now, while I still sometimes have pangs of guilt, I am generally comfortable with a low-level addiction to booze.

Apart from what we take into the body is what we put out to keep it going. And for health, that is exercise, natural for children and to a lesser extent for young adults, but harder for oldsters. Among our closest relatives, the monkeys and apes, the young play or eat during most of their waking hours. The adults spend plenty of time eating but do not waste much energy on play. Except for an occasional sexual or defensive act, when they are not feeding, they are resting. One cannot discuss their old age activities since they rarely have such, at least not in the wild. *Homo sapiens* is the only primate which really has an old age, mainly because he keeps his health longer. When monkeys get old, they die.

But the cultural primate, man, lives far beyond the years allotted to a chimp or gorilla because he has medical services, pensions, various mechanical and prosthetic devices and all the paraphernalia of culture. So he has evolved a most non-monkey kind of activity, play, or recreation, for his later years. Humans are supposed to look forward to retirement because then they can devote themselves totally to play. In the well-to-do cultures they can play golf, tennis, or go skiing, or simply take their RV's to various places to idle around with.

This isn't to say that older humans are up and at it like the kids or young adults. Not too many oldsters go in for touch football, basketball or volleyball. Walking, golfing and bowling are more their speed.

But the later years are supposed to be filled with fun. In fact, only the pre-school and retirement periods are expected to be totally for pleasure, again only in the well-to-do cultures. In tribal and many non-Western societies the small child and older person are supposed to do what they can productively. In fact, the release of the child from early, and often-times arduous, work is even recent for the Western countries. The European child was used as a factory worker well into the 19th century. In tribal and peasant societies children and oldsters were usually supposed to do what their bodies permitted. The old man in tribal society was supposed to make weapons and other tools useful for the hunt, while the old lady could weave baskets or cloth. In village India the old lady often kept the spinning wheel going, while the old man was arranging a wedding or some other social function, a time-consuming effort the way they did it.

But if all goes according to plan, we in the affluent societies need not work to help keep things going. We can, in fact, spend our time in the simplest of non-productive activities, we can "just sit." Some of us do. We have an old song, "The Old Rocking Chair's Got Me," referring to the fact that now being old, we can just stay put, even if gently moving. Moreover, unlike the not so very old days, such as those of my grandparents, we are not even condemned to simply sit and watch what comes by. Nowadays, we can watch television while sitting, we can become "couch potatoes," young or old. Of course there was radio in my parents' day, but so far as my memory holds, it did not have so much hold on sitters of any age. Now, about all that prevents our oldsters from spending considerable time in front of the tube is very poor eyesight.

But this is hardly exercise. Furthermore, I think there is little doubt that keeping active prevents the body from deteriorating quite so fast. But it's got to be something more than "sitting," with or without the electronic machine. For better or worse, it will have to be something in which the body gets up and moves.

Unfortunately, most pure exercise is boring. Furthermore, *Homo sapiens* is a creature highly subject to boredom. To go back to the other mammals again, the normal thing for an adult to do, when the stomach is full, there is no immediate fear of predators, no sexual drive (infrequent in any event) and no territorial harassment (also infrequent), is to rest. The lion which has just filled its stomach with meat, and the chimp which has just gorged on wild figs, will generally find a shady place to take a snooze. And though *H. sapiens* is not averse to a nap on occasion, this will cease before too long for most. The eyelids will pop open and the biped will soon feel a need to do something.

I know well. After an early lunch and reading for 30-40 minutes, I usually take a nap. But I wake up in 30-40 minutes, when I read again for a short while but then get up to type or do some farm chore.

Man is more prone to being bored than any animal I can think of. In fact, I suspect much of the creative drive among humans is merely a need to dispel boredom. I wrote in an earlier essay about how a group of men in my old neighborhood whittled down a house. Being out of work during the Depression, they had nothing to do. But instead of sleeping or otherwise resting, they gathered together on the porch of a vacant house and proceeded to whittle the boards into shavings, while talking and chewing tobacco. It was such a human scene, impossible to imagine among any other animals.

But boredom doesn't come merely from doing nothing. There are many activities that are boring, mainly because they are repetitious. One of the main criticisms of assembly line work has been that it is so boring. Anyway, any activity can become boring to humans; and this certainly includes exercise. The basketball or tennis player doesn't have this problem because his main goal is to win. The exercise is merely an added benefit.

But what about activities that have no other function than to provide exercise. Jogging has to be tiresome, especially if the runner continues on the same route over and over. Swimming a number of laps is boring. I once did such and continued only by concentrating on the count. But going back and forth all those times, watching the same skyline and pool boundaries, was boring.

I had a back injury years ago for which I was given a set of exercises to do "the rest of my life." I did them for many years through sheer determination. They were not so hard, but they certainly were boring. Then I found out that taking a 15 mile bicycle trip every five days would produce the same result. I quit the exercises.

A surgeon once recommended that after hip surgery I should begin exercises with a bicycle machine. I didn't do this because pedaling a bicycle in one place for a prechosen number of times has always seemed to me to be depressingly boring. I did get on my street bicycle to begin real outdoor cycling though, because I have found this less boring. The scenery varies and also I have developed a habit of noodling to pass the time while keeping my bicycle rolling. I frequently work out personal problems while bicycling. In fact, I wrote a book once which I entitled "Ten-speed" and subtitled it "The Thinking Machine."

With enough determination, one will continue a boring exercise. But without such, one might quit because besides being boring, exercise is frequently hard.

So what should less determinate oldsters do? My recommendation is to try to do those exercises which are least boring. Obviously games are good because one then has the added incentive of trying to win. The major limitation is that the old body is less capable than the younger one. So hard contact sports will hardly do. Obviously strenuous sports are okay if the oldster is super fit. I took up skiing a few years ago for the first time and had trouble learning. After letting me fumble around with a beginning group for a few days, the people at the ski school figured I wasn't going to make it. They were generous, however, and provided me with a personal instructor. I estimate that he was 5-10 years older than me. His wrinkled face and hands were well covered with liver spots. But there was no lack in his fitness. He could do pretty much what the fellows in their twenties could. He was also a good, patient teacher and in two days had got me through the basics. Later I learned that he had been skiing most of his adult life.

However, most oldsters will hardly be this fit, which is probably why golf is their most popular game. And though it doesn't provide much exercise, there is a little walking around in pleasant surroundings, as well as a few arm swings. Unfortunately, in their efforts to make the game easy, the industry has developed a means of eliminating the walking, the golf cart. Thus, more oldsters will get cardiacs on the golf course. In any event, it seems to be a pleasurable activity to many, more strenuous than cards or chess, but hardly something which gets the blood coursing.

So what's left for the over-the-hill bunch? The best life-enhancing exercises are those called aerobic, ones that get the circulatory system into action. Apart from being boring, running or jogging are the most frequent ones, though they are hard on the joints. Pounding on city streets is hard enough for a quadruped, but worse on an erect biped. I suspect my period of jogging contributed to the weakening of my hip joints, which required replacement. Still there are people who manage jogging without harm. I had a colleague who was 4-5 years younger than me and who had been jogging without letup for 20-25 years. He was in very good shape. But for most, the uprightness of *H. sapiens* makes him particularly susceptible to bone shock.

There are other aerobic exercises which do not have so much built-in risk. Undoubtedly swimming is one, since the body is mostly supported by water while moving. Then there is bicycling, my favorite, and again one in which the body is supported while the exercise is going on. And it can be as strenuous as the biker wants. Biking is also good in that the exerciser can go somewhere. For a number of years I went back and forth to work, ten miles one way, every day, and after I moved further away, I went one day a week, 20-25 miles one way. My most extensive bike riding was in my 50's when I did quite a few cross-country trips around California of 100-200 miles, usually with my son. My giant trip was in my mid-50's when I went from Santa Monica to Washington, D.C. But of course during those years I could still think of myself as middle-aged.

However, this was by no means the end. My bicycle served as an important therapeutic device after two major illnesses. The first was the stroke which I had at age 59, after which I used the bike to rebuild my strength and balance. I decided I had recovered fully the day I did a 35 mile trip and climbed to the top of Topanga Canyon on my ten-speed. The second has brought me near the present when I got back on the bike after surgical replacement of both hips. Again, the ten-speed brought me back toward my old strength and vigor .

Probably the most frequent aerobic exercise of our day is walking or hiking with a number of modifications to make it more strenuous. But simple walking is good in itself and ideally suited for the olding, since the pounding that comes with running is not a problem, and the boredom problem can be taken care of in a variety of ways. This is undoubtedly the best exercise for the getting-on bunch. I have a sister who continuously hikes with her oldster club. They get to see new places and talk to one another while doing it. It must seem like play, even if there is a considerable amount of exercise. They even take tours overseas in which walking is built in.

There are other less frequent kinds of exercises, some combined with play. Dancing is one, which is excellent if one knows how to do it. Some, like square dancing, are quite aerobic. Some years ago I joined a jazzercise group which had plenty of exercise.

And so to sum up, in the best of possible worlds we now have the oldster not eating too much, and with enough variety, besides doing some kind of exercise, hopefully to some degree aerobic. What next for health?

There was a time when Western man thought of his body and mind as two entities. We still maintain part of this fiction in separating mind doctors (psychiatrists) from body doctors (all the others).

The tribal medicine man did not separate body and mind in either diagnosis or treatment. And nowadays modern man is coming back to this way of thinking. The mind is simply a part of the body. Thus, it makes sense to think of the life process as one guided as much by the mind-body as by the body-mind, or either alone.

We now know that if the body is not doing well, that the mind will be affected. There are a whole series of body conditions controlled by the mind. One of the simplest, yet serious of conditions, is depression, a state that most have had to endure sometime during their lives.

My most serious experience in this respect was the depression following the stroke. As recounted before (The Stroke) I had already decided in the hospital emergency room that I was going to recover. I had no real basis then but I had the determination. I started to recover, then had an accident, went back to the hospital, and saw my marriage start to disintegrate. Depression set in, I stopped eating, lost a lot of weight, and was put back into the hospital. When I came back this time, however, a change took place. I decided to get well. I don't know what caused it exactly except perhaps a reoccurrence of the same determination I had managed just after the stroke. In any event, I started eating again and began to gain weight. I am glad I had my wife take pictures of me when I was at my lowest weight. I looked like a recently released concentration camp victim, all bony joints and an outsize head. And so I will always know what the mind can do to the body.

Obviously, we all want to avoid conditions of illness brought on by our minds, even though this is much easier said than done. The poor anorexic is frequently aware of her condition but still with all outside help, finds it difficult to begin eating again.

But here I want to turn to the more positive side of the mind-body syndrome. That is, while we can think ourselves into real physical difficulties, we can also think ourselves into better states of health. This is not a brilliantly new idea, though I suspect many of the ramifications have not been followed through, and particularly as it affects the olding.

The one resource older persons have is their thinking ability, simply because they have had so much experience at it and the mind does not seem to deteriorate so fast if it is used. Other parts of the body wear out faster, as any athlete can well testify. But despite jokes about senility and mental diseases of later years, the mind continues to work with age. The heaviest users of the body, those in active sports, generally have to quit in their 30's and 40's. But given the will, those who have depended primarily on their minds, will just be getting started. Lawyers, doctors, teachers, writers and business executives can keep going well into their sixties and seventies if they are not forced out by their employment system. The self-employed usually do continue long.

In non-Western cultures, appreciation of the wisdom of age has been institutionalized. The sage is by definition an older man. It has only been in relatively recent times, in Western industrial culture, that the wisdom of the older has been downplayed. Westerners have become such specialized individualists that they generally do not seek help from others unless they too are specialists. Moreover, because of rapid cultural change, it is generally assumed that the older are out-moded.

But the older person who is in relatively good health has a powerful tool which can be used for the good, productive life, his experienced mind. And so he should use the power of positive thinking. This is also an idea which has been around for some time, but most frequently in the past has been thought appropriate primarily for younger minds.

What is this power? There is nothing magical about it, and we know little about its physiology, but few medical people doubt its force. It is practically a medical rule that unless a patient wills it, he will not get well. This is not to say that will alone is always sufficient, but in most illnesses it is a basic requirement. Even some conditions which were formerly thought to be caused exclusively by external factors, like cancer, are now believed to be subject to mind force to some degree. The wife of one of my former colleagues was a therapist who claimed that the main hurdle in helping stroke victims is to get them to take a positive attitude. Otherwise, she claimed, her suggestions and procedures helped little.

The older individual can use this power, for helping to heal himself, but just as importantly to help create a good life. What can he do in particular? I think above all else he should look at the world with curiosity. He should be forever open and learning. I can think of nothing which keeps the mind more active, and thus healthy. However, this characteristic is also easier described than achieved, particularly with the olding. There is a natural tendency for the mature to become convinced that they have already learned what needed to be known. Too many try to maintain old habits, whether of food or politics. Unfortunately, the world does turn, and rapidly, and trying to hold back the clock does condemn many to being outmoded. In modern society things are not the same as they were when the oldster was growing up.

Like many younger people, when I was a boy, I was much interested in the unfolding universe. There was so much to know. There still is, and I am so happy for that. I was particularly fascinated with the wild world, even the man-degraded part that I had access to. I lived on one side of Indianapolis, Indiana, frequenting as much as I could the dumps, polluted river, railroad sidings and gravel pits. But I looked farther too when I got the opportunity. And the most important discovery I made to extend my universe, then and since, was the printed word. I read voraciously about animals, and later people, from places I could not then imagine ever going to. How I would have survived my childhood without books I cannot imagine. And in books I learned of many realms of knowledge beyond life forms. And somewhere I learned that the paramount searchers of the natural world made up a breed called scientists. Whatever else they were, they were endlessly curious.

Like many others growing up in America, in my late boyhood I faced the problem of whether to opt for a system of thought which was open or one which was closed. And though there are many other differences between religion and science, the most critical one to me has always been their relative openness. A revealed religion was a closed system about which there was little to discover other than what the prophet/gods had decreed. The basic truths had long been revealed and even what might be learned of nature was interpreted as manifestations of the revealed religion's reality. On the other hand, science was totally open. In science even the current beliefs were only temporary, to be accepted only until better theories came along.

The current brouhaha about teaching religious truths alongside evolutionism in the lower school system is consistently muddied by the advocates, and repeated by journalists. They argue that Darwinism should only be taught as a theory, not as a fact, unless religion is also. This makes the claim ridiculous, since Darwinism is only taught by its proponents as no more than the best theory available, to be replaced when a better one comes along. All science is taught this way. There are no final beliefs, as there are in revealed religion.

Anyway, back to my own growing up, it gave me a wonderful feeling of freedom to realize that there was so much to know and so little already known with the scientific way. How curiosity could thrive. Needless to say, I became totally dedicated to the scientific way and lost my conviction in revealed religion.

As I related earlier (On Becoming a Professor), one of the best student types we get in college nowadays is the older person who has come back. Most young people tend to come to college to get job training. The idea that learning will make a more human life generally is much further down the list (Introduction).

But one type nowadays who gets satisfaction from learning for its own sake is the older person, either retired, partially retired, or a part-timer, supporting him/herself with another job. And then there is the middle-aged or older woman who came back to college. Most of the above have genuinely relished the opening of their world that anthropology, and many other subjects, created.

I am describing science as my way to satisfy curiosity. However, there are other ways. The world of sight and sound have been explored by artists and musicians, and there too the possibilities have been endless for the curious. Stravinsky is as far away from

Beethoven as Jackson Pollock is from Michelangelo. All I am trying to get across is that the healthy mind is active and the fundamental need for activity is curiosity.

My most crucial discovery was anthropology. By the time I truly got into college, in my early twenties, I had shifted my main interest from wild animals to the ultimate animal, man. And when I learned that there already existed a body of knowledge which specialized in studying the diversity of this creature's way, I was totally hooked. Where else could one go to learn about the endless variety of customs that the cultural animal had evolved? And I learned that despite the large number of esoteric behaviors that had been discovered, anthropology had only scratched the surface. It was very clear that a veritable universe of knowledge remained to be discovered.

What next, do we let the mind at peace once we have got it stimulated to be curious? Sorry, no peace for the healthy mind. It has got to be busy. A vacant or lazy mind is not a healthy mind. But the mind of our hypothetical oldster is already busy learning about the universe. Okay, that's fine, but there's a further distance to go. All this new knowledge has got to be used. Just recording it won't do. The older person has got to be creative to be healthy.

More than any other animal, man has a tendency to get bored. Doing nothing is not good for the talking biped. One cannot be content leading a dog's life, a well-fed, well-cared-for dog, that is. The dog, as well as most other mammals, enjoys nothing more than lying around. This does not happen with man. Some lying around is okay. But sooner or later, mostly sooner, *H. sapiens* is up looking for something to do.

This starts very early in life. The oldsters whose memories permit must remember that the most doleful complaint of childhood was "There's nothing to do."

Of course there was everything to do, but with the inexperience of the early years, the child did not know what there was to do. Nothing much changed through the adult years except that individuals were given jobs. Many adults got tired of their work jobs and in their fantasies they replaced them with play jobs.

And then we come to those who have lived beyond their work jobs, the retired. What will they do -- sit? I've already described how if they want good mental health, they will keep learning with open minds. But then they must create. Does this mean that everyone past sixty has got to get busy on the great painting, sculpture, novel, symphony or scientific theory? Not hardly. To get into the league of the earth-shaking creators requires quite a bit more than simply the will of a person intent on creating the good life after sixty. But this does not mean that the normal oldster can do nothing.

What is a creation, anyway? It can be nothing more than a recombination of existing elements, or simply something quite common brought toward excellence. A fantastic blueberry pie is a creation. So is an attractive flower garden. A marvelous window garden of African violets is a creation. The greenhouse built by my construction neighbor is a creation. Jessica created a special personality in an Amazon parrot. The tropical fish tank I have at Hilltop is a creation. The possibilities are endless. Furthermore, there is no need of thinking about Michelangelo, Beethoven or Darwin. We lesser persons can also be creative in what we do. It gives us something to look forward to, but above all else it keeps our minds active, and thus healthy.

I delight in cooking innovative dishes. To put food items together to make a new combination is the only way I can enjoy this work. It thus becomes play-work, but nonetheless creative. I simply cannot cook by recipes. I don't mind reading them to get the general idea, but the combination I put together is always my own creation. Not infrequently people ask me for a recipe. Invariably I tell them that there is none, that what they just ate had never been made before, that it was a unique creation. This may not be the way to disseminate standard dishes, but that is not my business. Let others spread standard procedures, I will create my own in cookery.

I also love to write, and particularly because a language is such an enormous body of units. The permutations of the thousands of possible words are literally endless. The only limit to creativity in writing is the mind capacity of the writer.

I must repeat, however, that personal creativity is possible in any field of thought and endeavor. Though it will not loom large in world affairs, the growing of a giant pumpkin is a true act of creativity, as is the knitting of a fine scarf. Both contribute greatly to the mental health of the creators.

Man is a social animal, ordinarily content only in the presence of others. We live most of our lives in social groups, the main ones being our families, those we were born into and those that we made up by getting married and having children. But we also mix in with other groups, both formally and informally. As we go through life we change some groups while the rest of the membership of most others changes.

Thus, we have relationships, contact with other individuals in these groups, which is important for our well-being. There are many statistical studies which show that individuals who are married live longer lives than singles. We do better together in most respects, the older person as much or more than any.

The ties with the family of our birth are weaker now because of the modern life style. In band, tribal and peasant society the olding ordinarily lived with the family of orientation throughout their lives, at least the men. Girls, particularly in peasant societies, usually went off to live in their husband's families. However, in modern urban society we more often go somewhere else to set up our families of procreation, frequently to another town or state. Our contacts with our parents, brothers and sisters, and more distant relatives, become fewer. Our primary relationships are rarely with the members of our family of orientation.

Instead, we concentrate on establishing our own relationships with persons we search for. Also instead of having things arranged for us by family members, we select and arrange to live with someone else on our own. We have to develop our relationship on our own. In fact, the wide use of the word, relationship, is quite recent. When I was young, we did not use the term though we did bandy the word "love" about. I think by that we ordinarily meant sex.

Conditions have changed some, though far from totally. Although a primary relationship, as we understand it today, fulfills a multitude of social needs, one which is always included is sex. Ideally, we get it from our partner of the same or opposite sex.

But what does love and sex have to do with the older person? Aren't they well past that interest? Surprisingly enough, not necessarily. One of the most important discoveries of my later years has been that sex goes on and on. If I had been asked when I was twenty how I would be acting when I was seventy, I probably would have smiled. At sixty-plus it must be over. And that smile would come when I was having no sex at all. Although young men and women undoubtedly have greater physical capabilities, their ability to get a partner to the right place at the right time greatly reduces the possibilities. I remember a comment made by a middle-aged man in a singles group after a long group discussion about how to make love successfully. He said, "All the procedures that have been described are interesting but the basic one, how you get a partner to the bed, has not been touched."

Anyway, believe it or not, the older person is capable of making love, if he/she gets a desirable partner to the bed.

Unfortunately, since the break-up of couples is common wherever people depend primarily on love to hold them together, older persons will frequently have had several relationships, both in and out of marriage. I have had five, three being marriages. Those who have been satisfied with one person are probably better off, but conditions being what they are, this gets to be less and less likely.

Anyway, whether our other/s has been one or many, he/she is important for our welfare. In other cultures, where wider family relationships are normal, one's spouse or live-in is less important. Each person has a network of people from whom he/she can get help or support. But where the nuclear family of husband, wife, and children is the norm, other relatives are less important. Furthermore, since we in the economically favored cultures have longer lives, we usually do not have our children with us in our later years. This makes the spouse or live-in even more important.

Conditions have changed a great deal in two or three generations in the U.S., the period when we transformed ourselves from being ruralites to being urbanites. One of my grandfathers was a farmer, the other a bricklayer in a small town. When they and their wives got old, they went to live with their children until the end of their lives. My father was the only one in his large family who settled in the city. There he met and married my mother. Their three daughters moved to their own homes in different neighborhoods when they married. I moved out of the state. When my parents got old, they took care of one another until my father died, and my mother lived the rest of her life alone, at first in her own home, and finally in a nursing home. My first wife's mother also lived alone for many years after her husband died. My other two wives were younger than me and their parents are still living together. I have always assumed that I would live with my spouse or alone in my later years and now there seems little doubt that the latter will happen. Though I get along well with my son, he lives 3,000 miles away and concentrates on his family of reproduction. And if family matters keep going in the same direction in urbanized societies, my son's generation will be even more separated from their children and other relatives.

There are of course other kinds of pair relationships such as gay and live-in (without the social sanction of marriage). All are important, especially as we get older.

There is also the relationships coming from a network of friends. Since we depend a lot less on blood (gene) relatives, many of us seem to try to make up the difference through buddy relationships. We select these ourselves, so theoretically they will be more compatible than gene relatives. We exchange support with them because we want to. Of course the weakness of buddy relationships comes from the fact that there is no built-in, biological, tie. While our relatives are ours for our lifetime, self-selected buddies can and are dumped more easily. Also buddy networks in Western society seem most generally to be female. Overall, women depend more than men on relationships, including those of buddies. Girls are raised to be concerned with interactions between people, while boys are supposed to be individualistic competitors. Women watch the "Soaps" and read stories of romance, while men watch the sports and read action stories. This difference has become a significant point of contention. As the female star has risen in the West in the last couple of hundred years, one of the main accusations about men has been that they are not sensitive enough about relationships.

Thus, older women will continue to depend more than older men on relationships, of family, other relatives or buddies. If the older men can handle it, they will probably be better off with good relationships also, although they do have more other resources.

Once when I was gone from Hilltop, I was summoned back with the news that all my tropical fish had died. I rushed back, ironically from a three day affair of a standard person-to-person relationship, a wedding. The best friend of Jessica had got married with the full ritual. Jessica's friend had been so excited she couldn't sleep for several days, as well as having female trouble. Despite some changes in the role of women, girls in American society are still taught early and intensely that they should get married and that the marriage ritual will be the most important relationship event of their life. All had gone well at the wedding and by Saturday noon she was a more relaxed married woman. The next morning I got the fateful call about the fish.

What is the connection between tropical fish and a marriage party? Very simple, they are both significant parts of relationships. In all sorts of cultures people collect non-human creatures and shower attention and amenities on them. They make them into pets. Even tribal people, whose primary relationship with animals was to eat them, would frequently keep special ones of the same breeds, or others, as pets, carrying them around fondly. One of the well-described tribal groups of South America, the Yanomamo, got a reputation for being a very fierce people. Even so, the same warrior who would spear a person from another tribelet, would worry about losing his emaciated dog in a trade deal. In urban society, we usually consider having a dog as a pet for our children and will even select our own housing accordingly. Jessica and I, separately and together, have bought housing four or five times so we would have a place for dogs. Moreover, we spend a lot of money on our pets, frequently treating them like people, even finally burying them in human-style graveyards. And though dogs are kept more often than other animals, the range for pets is as wide as the animal kingdom. Some people and some cultures will make pets out of almost everything. The Chinese are fond of crickets and songbirds, tribal South Americans kept monkeys and macaws, East Indians have bears and gazelles and mongoose, while Westerners keep everything from boa constrictors to pigs as pets.

What do we get from these creatures? Basically, a pet is some animal which responds to us, if only to take food, while we get a creature on which we can lavish our affection and care. Some animals, mainly mammals, even reciprocate with affection. Others, such as reptiles and fish, do not usually establish strong bonds with us, though they do respond to care which makes them more comfortable. And they give us other things.

Formerly when I came into the living room, I would look first at the brilliantly colored fish in the salt water tank. Then the fish died, and looking at the tank made me sad.

It is claimed nowadays that animals calm us down and make us happier. Watching aquarium fish has been reported to lower the blood pressure of the viewer and thus make him healthier. When I explained my depression about the dead fish, my neighbor said, "I know how you feel. Pets become more important than people sometimes. I know I had to put a couple of old dogs to sleep and it was hard."

So there we are, a species which has evolved into a dweller of cities, usually surrounded by people, but doing all we can to keep other forms of life. Perhaps the reason we depend so much more on other animals is that we have almost full control of the relationship. They need us to survive and we need them on which to bestow our affection.

And what about the older person? Most of us concentrate on pets during our childhood, perhaps because we have great need of affectionate relationships then. We often find it difficult to understand and adjust to the adult human world. So we turn to our dogs, pigeons, turtles or crickets. They will give us love, or something that approximates it, if only we will feed and house them, and sometimes stroke them. Then we become adults and householders, with our work, human family and much else to keep us busy. If we get pets then, it is usually for the children. However, as we keep aging, our children grow up, and in urban society they usually go away. What to do then for an affectionate relationship? Most of us still have our spouse, but that relationship has usually lost some of its earlier passion. We still give affection, though it is somewhat attenuated, if for no other reason than that there is usually less or no sex. We can, and do, then turn to furred, feathered or scaled creatures for giving and receiving affection. The little old lady or gentleman taking the poodle or dachshund out for a walk is a familiar sight. Old Chinese men gather at special tea houses with their caged songbirds. The South American tribal plays with the monkey he did not eat.

And as usual, I turn back to my own history to describe what happens in a lifetime. My relationships with other humans were very tenuous when I was a boy. So I gave most affection to the animals I kept. Among them were a savage young alligator that had been given to me (in the days when there was a pet trade in these reptiles), a flightless crow which had been shot down by a hunter, pigeons I had caught, koi (decorative carp), chickens and rabbits, plus several dogs and cats. They varied in how much affection they returned and some fulfilled other functions also. The rabbits and chickens were raised for food by my father though several served as my pets. My parents had grown up in rural areas where most animals were kept for eating. I was bothered by what happened to some of the pets, but did what my parents told me to do.

When I grew up and became a householder with wife and child, I stopped keeping animals, except for several cats. My work took me overseas several times. Taking animals along was very difficult. In the place where I stayed longest, Laos, I built up a little menagerie which included some wild animals. Then I got divorced and my grown son went away. I started keeping animals again and have increased their numbers steadily for the past seventeen years, even though I married again twice.

My first wife was not heavy into animals, though she was willing to have cats while our son was growing up. My second wife remembered one dog she had had as a girl and we got another spaniel for her. Also I kept a tank of tropical fish. My third wife remembered her pet dogs, birds and turtles fondly when she too was having considerable problems adjusting to the adult world.

Now my son is a householder with no animals. He tried a dog for his children, but it was too difficult. He spends most of his time in the world of work to maintain his human family.

Anyway, older people will do well keeping pets, particularly in the modern, urban world, where human relationships are more difficult to keep going. Before the urban revolution there was a simple division of labor, men doing muscular work while women did lighter tasks. Children did the same things as adults, though scaled down to their lesser capacity. All that changed as society evolved an ever greater division of labor, based on formal education. Moreover, social change steadily accelerated. Thus there came a day after the industrial revolution when one's work life was strictly a matter of what one had learned. Older persons were no longer valued for their greater, though out-dated, experience. What they had learned was frequently thought of as outmoded. Who would go to an older person to learn about computers, rocketry or robotics? These were the fields of the future which changed so fast that only the young and middle-aged could expect to be current.

Even so, after six or seven decades, the olding have accumulated much experience in the workaday world and elsewhere. Unfortunately, because of other needs, many have not done what they wanted to. After a lifetime of work in a field where their freedom of action was severely curtailed, or because of other restraints, many have looked forward to the day when they could discontinue, to retire absolutely.

Lucky indeed is the person who in modern society has been wholly content in his field of work. There are some, though they are a distinct minority. Those who have achieved a high level in their field of endeavor are usually content, whether it is in football or music or politics. After all, besides getting well rewarded materially for their efforts, they also got public acclaim. Also those who have had a degree of independence are usually more content than those who have been heavily supervised. If not in dire economic straits, the general farmer before the era of agro-business was usually a satisfied person. Also the college professor, who planned his own courses, was normally more content in his field than the primary school teacher, who has had to accept the dictates of the school system. Also, those who have been connected in vital ways to the ongoing affairs of mankind or the world, the ecologist, plant breeder, forester or city planner, will usually have found great satisfaction in their work. Then too, the scientist and artist have usually been satisfied, since they got to use their creative faculties.

Despite having had to go through periods in my work career when I had to accept considerable constraints, and even had to do some unpleasant jobs, overall I had considerable freedom to do what I wanted. Furthermore, I could use some creative talents and was in a field of science where curiosity was well rewarded. In my teaching, I was mostly allowed to do what I wanted. Thus, I came to a period of semi-retirement with no desire whatsoever to leave my workaday field behind. I was content to know that it would be anthropology all the way. Retirement would simply be a period for exploiting even more my commitment to anthropology, even if in different ways. But I would continue the old ways as long as possible also, even if with less intensity. Although I have decreased the amount of teaching by two-thirds, I will go on with this activity as long as possible. I have laughingly explained that I will not leave the lectern before I am dragged off unconscious. The continuity between my work life and retirement couldn't have been greater. This is perhaps the closest I, and others like me, can come to the idea of earlier and simpler societies when old meant wiser. Old men were then sought out for their accumulated wisdom. Now this privilege is no longer given automatically. Older in modern, urban society more often means out-of-date. And yet if one has continued to change with the times, and speaks with the authority of accumulated knowledge, he may still be listened to.

What I am saying in essence is that the olding will be better off using the knowledge of their experience, only being careful that they do not enshrine what was, instead accepting the world as the rapidly changing scene it has become. Furthermore, if they have been fortunate enough to have continued in the same field for most of a lifetime, they will do well to exploit it in later years. If not, they need to exploit whatever experience they have had.

The great majority of societies have depended heavily on the powers of the supernatural. There have been legions of such systems, known collectively as religions. When man could not deal with the real world to his satisfaction, he frequently turned to heavenly powers. He prayed and made offerings for having children, for better health, for success in this world, for power, and for much more. He also extended this world into both a before and an after. The afterworld was a place where he would go, once the travails of this world were over. Not surprisingly then, the older members of society have in general paid more attention to getting squared up with the supernatural than have younger people. Youth, after all, could more reasonably ignore the end, to live with the illusion that this life would go on forever. But aging has always carried the certainty that life would end. Thus, older persons have devoted more time to propitiating the supernatural.

In Hindu society there are four stages to a person's life - child, student, adult householder and elder. In this last, the older person was expected to devote much attention to the search for salvation. When I did field work in India my assistant was just between being a student and a married householder. When I came back the second time, over twenty years later, his children were grown, he was considering retirement, and had installed a holy man in a separate room of his house where he could get help in seeking the infinite.

When I was a church-goer in my youth, older persons dominated in numbers most services during the week. And though I can recognize the need for the olding in general to spend more time and energy in preparing for the afterlife, I cannot personally take part. The trouble is that I early lost conviction in the reasonableness of any of the multitudinous afterworlds of man's cultures. To me they all became wish projections, validated by group acceptance. It became impossible for me to believe that any force capable of managing the universe would also be concerned with the minuteness of human existence. The only supernatural system which seemed at all reasonable was original Buddhism, without the many accretions of later history. And even if I never got around to accepting that system either, I felt that it was more reasonable than the others, simply because man was figured in it as no more important than any other entity.

So I cannot suggest what would be the most desirable way for the olding to go in this regard. Simply because so many do exert themselves can just as easily be explained on the grounds of need as any other. Older persons, being closer to the end, do have more need to square themselves with the supernatural, so long as they have faith. If they do not, as I do not, it is another matter.

On the other hand, there is an advantage in not being concerned with an afterlife. Then one can continue, as long as there is sufficient energy, to concentrate on livingness, on the now of doing, about which there is far less doubt of reality. That has been my road, and I have no regrets.

INDEX

A
Aboriginal, xii, 261
Accounting, 29
Adventures, 263
Advertising, 187
Aerobic exercises, 292
Age, 53
Agency for Inter. Develop., 58,68
Agnostic, 219
Agrarian, 185
Agriculture, 101,149
Air Corps, 259
Alcohol, 287
Aldine Publ. Co., 71
Amer. Anthro. Assoc., 54, 68
Amerindians, 1,7,175,261
Anglo-American, 172
Anglo-Saxon, 201
Animals, 88,129,140
Anthropocentrism, 114
Anthropomorphism, 114,171
Antlropology, 25,29,244,297
Ants, 225
Appearance, 184,191
Arranged marriage, 185
Araucans, 111,195
Arensberg, 35,68
Army, 253
Avocados, 88,148,154,156
Aztecs, 161,256

B
Baboons, 183
Baloney, 153,
Bard, 237
Bees, 203
Behavior, ii
Benedict, Ruth, 37
Bestiality, 199
Bicycle, 156,248,250,281,291
Biology, 47
Birds, 206
Blood Pressure, 246
Blue Highways, 91
Bobcat, 226
Boggs, Steve, 69
Book, 78
Bombing, iv
Boomers, xiv
Boredom, x,46,289
Brain washing, 210,265
Breeding, 134
Bull, 197
Burn-out, 45

C
Causes, 137
California, 60,155,240,286
Calif. St. Univ., 62
Camp Pendleton, 264
Cats, 197
Catholicism, 4,80,82,218
Centrism, 164
Chairman, 62
Change, 103
Chardin, ii
Chauvinism, 47
Chickens, 111,195
Children, 233
Cherimoya, 204
Christianity, 19,169
Church, 305
Cities, 187,229
Civil service, 56
Civilizations, 10,177

Index

Cleveland Nat. For., 91
Cobras, 144
Cockroaches, 122
College students, vii,xii,xiv
Columbia University, 55
Communication, 233
Computers, 188
Communists, 261
Constipation, 248, 285
Conversion, 169
Cooling, 112, 298
Cooperation, 233
Copulation, 201
Coyotes, 93,115
Couch potatoes, 289
Culture, 15,78,181,183
Creationism, 4,7
Crash, 263
Creativity, 278,298
Credit, 274
Crevasse, 240
Culver City, 95,106,109
Cultivators, 283

D
Darwinism, vi, 5,39
Dating, 156,186,234,286
Depression, viii,277,248,293
Deceit, 232
Desert, 127
Dieting, 280
Diffusion, 17
Dictionary, 39
Discharge, 260
Divorce, 92,158
Dizziness, 251
Doberman, 151,209
Dogs, 197,208

Domestic animals, 199,227
Domestication, 101,179
Dreamtime, xi
Dress, 34
Drugs, 286
Duality, 29
Ducks, 206

E
Earthquake, 240
Earth Diver, 7
East Asian, 168
Eastwood, Clint, 165
Eating, 249
Ecology, 103
Economics, 158
Egocentrism, 163
Emigrants, 186
Enculturation, v,28,256
Endangered species, 181
Entertainment, 235, 289
Environment, 223
Equality, 24
Ethnicity, 171
Ethnocentrism, 6,10,33,171,210
Escondido, 84
Evolutionism, 7,8
Explanation, 152
Examination, 55
Exercise, 288
Exotica, 52
Extinction, 104

F
Familism, 27,31,167
Family, 105,300
Fallbrook, 243
Fantasy, 128

Far East, 168
Father, 253
Feminist, 47
Fertility ritual, 160
Fiction, 73
Finances, ix,273
Finches, 134
Foods, v,17,34,153,279
Forest destruction, 103
Four letter words, 201,205
Folk tales, 79
Frazer, Sir James, 162
Fried, Morton, 36
Fulbright, 56
Function, 19,235
Funeral, 254

G

Guatemala, 149
Garden, 112
Gender, 178
Geneva, 144
German, 254
Generation X, xiv
Gold market, 275
Golden Bough, 159
Golden Years, 267
Golf, 291
Gossip, 34
Grass clippings, 109
Greek, 29,44,257
Group, 166
Gull, 120
Gutenberg, 78

H

Health, 277
Hiking, 292

Hill, 87
Hilltop, 202
Hindus, 146
Hiring, 62
Homocentrism, 180
Honorific, 268
Holistic, 20
Homo sapiens, ii,iv,130
Houses, 87
How-to, xi
Hucksters, 272
Human rights, 173
HumRRO, 60,68
Hunters & Gatherers, 283
Husbandry, 101
Hyacinths, water, 108,135

I

Ideas, iim,vi,254,295
Inalienable rights, 174
Inarticulate minority, 46
India, 139
Indiana, 196
Indianapolis, xi,24,82
Individualism, 16,165
Industrialization, 167
Influence, 234
Information, 211
Innovator, 165
Insects, 121
Inter-relationship, 19
Interviewing, 213
Introducing Social Change, 71
Investment, 274
Islam, 169,177

Index

J
Jack Russell terriers, 208
Jains, 146,224
Jane Goodall, 212
Jews, 14,169
Jessica, 150,204,214,224
Jobs, viii,139
Jogging, 291
Jumbies, 213
Jungle fowl, 141

K
Kentuckians, xi
Killing, 224
Kinship, 6,167
Koi, 108,207

L
Language, iii,21,79,161,212
Laos, 59,140
Last Emperor, 36
Latinos, 148
Lawrence of Arabia, 37
Learning, 184
Leftovers, 107
Lenin, 236
Liberalism, 173
Library Journal, 84
Librarians, 81
Life, 220, 223
Listings, 191
Loan service, 86
Lore, 68
Loris, slow, 141
Los Angeles, 116
Love affair, 203
Lunch, 150
Lying, 238

M
Machismo, 180
Magic, 73,160,214
Manchus, 36
Malaysia, 152
Mana, 220
Marital ads, 57
Marriage, 157, 193
Marsupials, 118
Mating, 196
McDonalds, 269
Media, 264
Medicine ,76, 243
Mega-wars, 265
Mexico, 149
Microscopic creatures, 123
Military service, 175
Milwaukee, 56,67,116
Mind altering, 286
Mind expansion, 91
Mind power, 50,294
Miranda decision, x
Miracle, 221
Missionary, 3
Mulch, 108
Museum, 67
Mobility, 168

N
Nation, 174
National Singles Reg., 188
Nazis, 173
Nehru, x
New Guinea, 48
New York, 116
Newscasters, 272
Nigeria, 143
Nineteenth century, 80

North American, v
Nudity, 96
Nul hypothesis, 66
Nurse, 28

O

Occupation, vii, 3, 27
Oddball, 49
Offense department, 258
Old age, 288
Older person, xiii
Omnivorousness, 119
Opossum, 117
Order, 106

P

Padre Padrone, 199
Pairing, 167
Paralysis, 246
Parochialism, 33
Peace Corps, xiii
Peanut butter, v
Peasants, 233
Personality, 191
Personal experience, 52
Pets, 117,139,227,301
Ph.D., 29,35
Pigeon, 121
Photo dating, 189
Plants, 128,202,203,227
Plough, 179
Pleasure, 191
Police, 28,44
Privileges, 179
Profess, 39
Professional, 54,81
Professor, 65
Progress, 165

Proxemics, 94
Prudishness, 198
Puberty rite, 48
Public libraries, 81
Public space, 97
Publications, 65,75
Publish or perish, 66

Q

Quadriplegic, 221

R

Rabbits, 115
Racism, 14
Rational, iii
Rationalization, 20
Rand, Ayn, 165
Reading, 79
Real & ideal, 23
Real estate, 275
Refinement, 285
Relationships, 192,299
Religions, 169,222
Reptiles, 207
Reproduction, 228
Relativity, 11
Retirees, 270,298
Revelation, iii
Road kill, 115
Robber barons, 166
Romance, 157,185
Rumpelstiltskin, 64

S

San Diego, 271
Santa Monica, 157
Scenery, 90
Science, 295

Senior citizen, 269
Sensitivity, 23
Sex, 198, 229, 262, 299
Sexual revolution, 200
Sexuality, 183
Sir Walter Raleigh, 18
Sheriff, 27
Size of animals, 120
Skiing, 291
Skin, 130
Skinner, ii
Skunk, 117
Slaughterhouse, 142
Slave market, 59
Snakes, 129
Social, 19, 22
Soldiery, 257
Solar energy, 133
Sorcery, 73
South Pasadena, 155
Space, 95
Sparrow, 121
Spice, 17, 151
Specialization, 118
Stalingrad, 255
Stewardess, 159
Stewart, Jimmy, 260
Story, 237
Stroke, 222, 278
Student, 27, 41
Sun burn, 130
Sun worship, 135, 220
Sunlight, 126
Super-national, 76
Supermarket, 282
Surgery, 277
Symbols, iii

T

Taiwanese, 28, 42
Talking, 4, 184
Tax, Sol, 70
Teacher, 40
Teaching, 250
Technical Change, 69
Technology, 110, 21, 102
Tempo of life, 133
Tenure, 66
Territorialism, 175
Therapy, 91
Thoreau, 165
Time capsule, 162
Tomatos, 250
Tobacco, 153
Tools, iv
Traffic, 18
Traffic cop, 268
Trash disposal, 103
Tribes, 32, 185
Trinidad, 74, 154, 213, 216
Tropics, 127, 1 55
Tuna sandwiches, 150
Turnbull, Colin, 164
Twain, Mark, 232

U

Uniforms, 260
United Nations, 70, 144
Uppers & downers, 286
Urbanization, 70, 144

Y

Yiddish, 171

GN
27
.N54
1998